Neural Networks are Homeomorphisms:
An Introduction to Higher Mathematics for Decision Scientists

Serge Plata

Bibliographic information about this book can be found at https://www.sergeplata.com

The graphic design of this book was done by Nelly Sugar and Clapham Publishing Services.

The illustration on the cover is a reproduction of a work entitled "Through Dreams" by Nelly Sugar. Nelly Sugar ©2021. For a complete reference: Sugar, N., (2021), *Through Dreams* [Oil on Wood], London

ISBN: 978-1-7392031-7-7
©Serge Plata, 2023

14 Burnett Square, Hertford, Hertfordshire, England

To Nelly

Contents

About the Author

Doctor Serge Plata obtained his PhD in the mathematics department at Imperial College London, he also studied an EMBA at Harvard Business School, he was awarded a fellowship at the Institute of Mathematics and its Applications in the UK. He is a chartered mathematician (Royal Charter) and a chartered scientist (UK Science Council) in the UK. He is author of *Visions of Applied Mathematics* published by Peter Lang and other papers, articles and editorials for specialized magazines. He has been keynote speaker in several data events and academic conferences in the UK and abroad. He is a certified six-sigma black belt and he has held directorship and leadership level roles in data and decision science, AI, analytics and R&D from FTSE 100 and S&P 500 companies to startups and also international management & technology consultancies.

Acknowledgements

The inspiration and development of this work came from a series of rather informal but seriously theoretical discussions on the subject over dinner and after-hours sessions with Alec Boere through the period of approximately two years, in which he suggested among other things, the need to develop a framework and foundation for AI and decision science that requires a general knowledge of advanced mathematics and its applications. I am hugely grateful to Alec for this and his constant input and support. In addition to Alec, I would like to thank all the people who put their personal own time even in weekends and after-hours to support me:

David Semach for his invaluable intellectual contribution and for showing me all different angles of our technology.

Debolina Guha Majumdar, for her endless innovative ideas, out-of-the-box suggestions, and for teaching me the strategic side of decision science.

Maxi Zattera for being so generous with his thoughts and creativity but above all for sharing his philosophical approach to data science and for supporting me unconditionally.

Jovana Markovic Ioannou, for giving me all those unique pointers on all subjects, emotional support and expounding the positive use of her critical mind.

Kofi Awuma for showing me how to synthesize technological practical applications.

Martin Weis for mentoring me and showing me the lighter (but serious) side of data and decision science and how to balance theory and applications with a laser-focused approach.

Daniel Scain for being my sounding board and for putting on the table so many ideas and concepts in the years.

Lubna Ahbedin for sharing her sharp multidisciplinary approach and her emotional intelligence.

Krzysztof Bisewski, for his incredible mathematical knowledge and true understanding of AI foundations.

Harald Gunia for his constant input with research references.

My brother Gerry Plata, Dr. Claudio Canaparo and Prof. Eduardo Ortiz† for everything through all the years.

A special mention goes to my editor Dr. Katie Isbester of Clapham Publishing Services and its book designer Petya Tsankova for sharing their expertise and skills and being instrumental in the publishing of this book.

But above all, I owe my wife Nelly a great debt of gratitude for her invaluable input to the design of this book, its artistic cover, and of course her love and unconditional support. She is an inspiration to be better every day. This work would not have been possible without her.

Introduction

> Geometry is the art of correct reasoning from incorrectly drawn figures.

Science and Method
Henri Poincaré

Firstly I would like to say that all drawings in the book are mine and are done in the same spirit as the words by Poincaré, so do not expect fancy drawing, but definitely good reasoning.

The naming convention, specially in the index, is a personal choice of mine following computer science rules. I use only lower case except for proper nouns. This is because even with known and well-defined naming conventions in computer science, it is extremely difficult to be consistent, generally creating confusion[1]. To avoid this, I decided to adopt a convention rule that is internally consistent.

---

[1]Examples of naming conventions are CamelCase, capitalize all nouns or use non capitalized prepositions and articles but all the other nouns can be (or not) capitalized, etc.

Another note on the theorem proofs is that they are really well thought out. I decided not to prove some theorems as the proof would not add anything to the purpose of the book, and moreover they could be found in classic textbooks.

In these cases, I will just refer to the text to find the proof. The proofs that I included add knowledge and are developed by me, for example out of them we can extract problem solving techniques or set foundations for practical modeling.

There are three major original contributions in this work, two of them are purely mathematical and the other one is philosophical as it addresses epistemological explanations and foundations through the application of two theories forged during the last two years:

1. The geometric interpretation of the rotation number under ergodic transformations. The classic ergodic theory studies only view this concept as an algebraic topic, but given my background in geometry and topology I created my own original and novel geometric view and study. Even absolute values under ergodic transformations are interpreted as distances, which eases the study of these special dynamical systems from a visual perspective. And although this is a good interpretation I do not leave aside the algebraic side of it, so I play with both sides iteratively.

2. The versatility on the approach when treating a fair amount of mathematical areas. Normally books specialize in one topic, be calculus, algebra or any other, and moreover the most specialized areas ones are even more concentrated in one and only one topic, for example set topology or group theory. I wrote this book covering

topics from geometry, topology, algebraic topology, measure theory, algebra, linear algebra, ergodic theory, spectral theory, group theory, calculus, functional analysis and dynamical systems.

3. A whole epistemological proposal as the application of two theories construct knowledge empirically but at the same time reconciles with the classic axiomatic mathematical methods, hence proposing a pedagogical perspective towards higher mathematics.

These three contributions make this book unique as it proposes new angles to construct mathematical knowledge and covers a wide variety of fields with reasonable expertise. The advanced reader will recognize the great depth with which I cover all the mentioned subjects above as none of them are covered shallowly.

When I started my studies in mathematics I always tried to find practical applications of the theory. And that is why I tried to understand physics, economics, operations research and in general any domain I could apply mathematics to.

Then, when I started working on data science and applying computational solutions to industry problems, I had to go the other way around: find the theory that supported the practical application and empirical results. This was certainly a different way to approach mathematical theories. Inevitably, since those years I had to apply a combination of analytics and theoretical models and this is what we call now "decision science".

The explosion of data in the first decade of this century led also to the fast proliferation of higher and complex models in computing for which higher mathematical tools and concepts are fundamental, not only for their understanding but also for their future development.

Today, thanks to AI and data science, we find a number of mathematical models embedded in solutions (mainly in the form of computational code) in finance, marketing, health sciences and basically in any human activity. But most importantly the application of those models which we call decision science, could not be possible without the mathematical knowledge and understanding of the models.

And although there is a practical side of it, which I learnt thanks to my mentors and other people who stimulate me intellectually and gave me the space to create, there is also another side, the abstract technical side.

The former is important to land those ideas pragmatically, but the latter is crucial to a) create more of these ideas b) to really understand what those models are about and can do and c) most importantly apply them to practical cases and help in the decision making process.

This book is about the b) and c) as it will allow decision scientists and other technical professionals like data scientists and computer scientists to get to the heart of the models and algorithms and launch more ideas and creative initiatives to grow the field and support business's decisions and processes more meaningfully.

The main objective of this work is to show those fundamental concepts in action. This will be done via the analysis of the spectra of ergodic transformations in Abelian compact groups. It sounds daunting, but by analyzing this we will go through all the basic concepts of higher mathematics usually found in the underlying theory of data science models, and AI research papers and the recommended actions based on those models e.g. operations research.

Although this sounds rather complex and would scare even the specialists in pure mathematics, the reader will realize that following a method (in this case trying to solve a problem) concepts will be a lot easier to understand and later apply them to specific situations and solutions.

Technically, I will mainly concentrate on the group of the circle S^1 and its rotations and for this, I will go through the necessary theory to understand the mathematical concepts. Why the circle and its rotations? because this example encompasses dynamics, statistics and continuous transformations; so almost all what you find in modern data science models. Also this topic covers concepts that are crucial for computational modeling, like eigenvectors and eigenvalues.

Moreover, this book should be considered as a general book more than a specialist one as it touches several mathematical fields, from topology to algebra, group theory algebraic topology, spectral and ergodic theories, linear algebra and advanced analysis including measure theory which serves as the foundation of statistical analysis and probability which are crucial for decision science.

Statistical analysis is closely connected to machine learning but also it is to differential topology as it is the foundation of neural networks. In fact we can define data science as the field of study that overlaps primarily mathematics with computer science to extract information and knowledge from data.

Decision science is the field of study that overlaps data science with analytics in order to make a decision. In this sense its task is to analyze foundational factors of choice and their context. Data science will get the normative side while analytics will work on the empirical one.

Please note that there is a difference between data (which is the plural of datum), information and knowledge. Data science deals with data and information and decision science deals with knowledge and its direct actions.

Neural networks are firstly function approximation algorithms. That is precisely where the power of the neural networks rely. If we have a function, then we can calculate maxima, minima, inflexion points, range, we can get predictions and define patterns, etc. Traditionally we could approximate data with functions via Lagrange polynomials, regression models and other methods, however neural networks are a great way to do this.

The idea of Neural networks started as early as the 1940's with the seminal paper by [McCulloch, 1943] where they proposed a model of a single neuron. Later [Rosenblatt, 1958], called it "perceptron" ans since then we have adopted this terminology to refer to artificial neural networks. Obviously these networks are nothing like the actual human neural networks, and they don't intent to mimic neural synapses, but they certainly have similarities specially on the "network" side.

So in general a neural network is a function approximation algorithm based on a combination of "nodes" called neurons in a network. These neurons perform operations in the form of functions. These functions are of two types: aggregation functions and activation functions. Normally an aggregation functions would be a linear combination of terms. The neurons are organized in layers and we have three types: a) the input layer, b) hidden layer or layers and c) output layer[2].

Neural networks, being function approximation algorithms, are in essence homeomorphisms as they transform and adjust *smoothly*.

Hence if we understand these concepts we would be able to unpack advanced papers and research and hence properly understand models and algorithms widely used in artificial intelligence.

When we say that neural networks are function approximation algorithms, a relevant question is: can they approximate any function? continuous or discrete? What are their limitations? etc. an adequate answer is that in order to perform back propagation, neural networks require continuity and differentiability. In other words, neural networks are in essence homeomorphisms (actually they are homotopies in a strict sense), otherwise it would be impossible to adapt at every epoch or iteration. Even when we apply gradient descent we are playing with derivatives which require continuity.

---

[2]This book is not about neural networks as such, but their mathematical foundations and other higher mathematics concepts behind the machine learning algorithms and modeling. Therefore I will only define them briefly so the reader gets familiar with the context.

Neural networks are similar to dynamical systems. They iterate, they are based on a given function (called the aggregation or activation function) and moreover, they have continuous transformations.

The space of continuous transformations is called a topology. A topology is where continuous functions "live", hence neural networks are closely related to differential topology.

And the chain goes on and on. The links from decision science to advanced fields in mathematics are endless; that is why the need of this book which encompasses many different methods and areas closely linked to the profession of data and decision science.

Through this book, I will show an easy and practical -nonetheless formal- way to look at topologies, homeomorphisms, and other mathematical concepts as they are the starting point to understand the essence of artificial intelligence algorithms.

Another aim of this work is to understand the mathematical foundations of advanced analytics and computing, mathematical modeling and real-life applications of higher mathematics and not only machine learning algorithms. This is essential to the normative side of the data-based decision making process.

After advancements in AI (like GPT3), data science will likely be absorbed by a machine. What the machine might struggle with, at least in the short term, is the evaluation of the contextual application (or decision science). That is why it is important to learn the first principles of technology which are based in mathematics.

This book contains serious foundations of mathematics and computing and would never propose ideas like "become a data scientist in 3 months", or "you don't need math to be a data scientist" which only trivialize and degrade the serious and complex world of data and decision science not to mention misleading people to make them think they can do some analysis without understanding the concepts, but by merely manipulating some off-the-shelf libraries and tools that normally would take years to understand. In the end these mechanical tasks will be done by machines and that side will be replaced by automation. Moreover, the people who created such libraries, who were the ones who actually understood the mathematics behind the models, are not likely to be replaced by machines.

The need for understanding the real issues behind the theory is fundamental in the problem-solving process. The "recipes" that many people apply sometimes do not even make sense to the problem they are trying to solve. In my more than 15 years' experience in the field as an active practitioner from start-ups to FTSE 100 and S&P 500 companies, I have witnessed the problems created by these practices and approach, that possibly not in a malicious way, but nonetheless have contaminated and damaged the discipline.

Hence, one of the main points of this book is to facilitate the understanding of concepts in the mathematical, computational and artificial intelligence fields and to have the tools to build complex applied modeling, design thinking and problem solving in real-life situations.

In other words, I will try to make a serious effort to dig into the heart of models by understanding the mathematical foundations to allow creativity and growth in the field as only in the abstract layer of mathematics we can see the many applications of the theory.

Along the text I will define concepts and mathematical objects precisely without following any text book in particular, which will allow a unique way to understand the main ideas and their formalization.

This will allow the reader to go further and be able to read research papers on their own. Therefore, the definitions will be formal and rigorous and -if not the same- they will be equivalent to any definition found in the classic mathematics texts.

Chapter 1

Foundations and Generalities

> In Alice in Wonderland the Red Queen tells Alice she has to run as fast as she can just to stay where she is.
>
> *The Lost World*
> *Michael Crichton*

We will start with general definitions because in a way, this will show how mathematics is written and will give the foundation to read mathematical and data science papers: In mathematics we normally have a definition, on which the following concepts will rely on; definitions are like axioms and do not need any proof but certainly agreement; once agreed, and after one or many well established definitions, theorems will be enunciated. Theorems are statements based on definitions that need proofs. However I will also iterate on concepts and construct mathematical ideas outside the classic axiomatic process.

It is important to note that I will not prove all the theorems as some of them are well known and can be found in any of the classic textbooks. For this I will refer to the source, however I will prove the theorems that involve some important analytical tools and-or applications of important theoretical concepts as in them will rely the key knowledge and aim of the book.

One of the main difficulties in mathematics is the dual nature of some concepts. For example when we talk about a vector we normally mean the two-dimensional Euclidean vector which is an ordered pair or a coordinate on a plane, and its representation as an arrow and even with that, sometimes we refer the vector as only the head of the arrow, sometimes the length of the arrow and sometimes the entire arrow; students have to pay close attention and discern in which case we are talking about which concept (even if we say *vector* all the time).

So firstly, and following the remark in the previous paragraph, I would like to outline the concept of a function. The concept has evolved historically: in 1668 the mathematician and physicist James Gregory defined it in his "Geometriae Pars Universalis" as *a quantity obtained from other quantities by a succession of algebraic operations or by any other operation imaginable.* This definition is correct, and it is the one with which we start studying them in our middle school courses.

But a function has other meanings too. A function is a transformation, but also a rule of correspondence, but also a law of change, but also a pattern, a behavior and a relationship. How about that for duality?

When we mention function , we might be referring to a mapping ; in fact a map (in the cartographic sense) is a function. A function describes the pattern and behavior of phenomena in nature. For example planetary movement, sales cycles for a retailer or the stock market.

In the XVII century many scientists like Leibniz, Newton, Kepler, Bernoulli, Mersene, Pascal, Brahe, Galileo, and many others, found functions that described the physical laws of the universe.

But why are functions related to neural networks? Because neural networks are function approximation algorithms. They take data, just as the people mentioned above, extract the pattern and create a function that describes that data.

Now you see the power of neural networks? Once we have the functions, we can calculate maxima, minima, cycles, and most importantly we can predict with a fair amount of certainty what will happen in the future.

When I was in school and the teacher asked me to answer a question like "sales are described by the following function... calculate the maximum sales using derivatives" I always thought, "how do you get the function in the first place?".

Well, in my work as a data and decision scientist I always tried to find those functions, the laws that rule certain phenomenon, be sales, inventories, or the stock market, which is not an easy task as many variables are involved. Fortunately computational solutions like neural networks helped us in some way by approximating the functions that describe behaviors. A good practical example is given by [Boere, 2021] where he explains the application of Monte Carlo tree search in banking.

I have to say that other models that we know are contained in the machine learning field try to do the same as the people named above. Gauβ and other scientists knew that phenomena in the universe were affected by many variables and that basically any problem was a multivariate one. So they tried to find the three most important ones, the three variables that accounted for say 90% of the phenomena. In mathematics we know this as variable reduction, in data science this is well known as dimensionality reduction.

Even the three-body problem was acknowledge by some of the pioneers in the field, but they knew that an approximation (with two bodies) was more than enough to continue with the scientific solution[1].

The three-body problem led to the study of chaotic dynamical systems. These systems aim to predict -among other things- how phenomena in the universe evolve in time, but nevertheless, we need to find a function.

1.1   First Formalization

Neural networks need "aggregation functions" and "activation functions" and dynamical systems also need functions. So this is a good point to start formalizing terms, as the aim of the book is to establish precise concepts more than vague notions.

---

[1]The three-body problem comes when trying to model the interaction of three particles. Newton's theory involves two masses, but when trying to model the interactions between the Sun, the Earth and the Moon, the system of equations was too complex. The two-body problem has a closed solution, but not the three-body. When involving three bodies, we need to get into chaotic dynamical systems which numerical methods and computational solutions are normally required.

Definition 1. *Ordered Pair. Let X be a set[2] and $a, b \in X$ we define the ordered pair formed by a and b, denoted by (a, b) by*

$$(a, b) = \{\{a\}, \{a, b\}\} \tag{1.1}$$

As we know, $(a, b) = (c, d)$ if and only if $a = c$ and $b = d$, and this definition will lead us to establish an order in relationships.$\{a,b\}$

Definition 2. *Cartesian Product. Let A and B be sets. The Cartesian product of A and B denoted by $A \times B$ is the set of ordered pairs:*

$$A \times B = \{(a, b) \mid a \in A, b \in B\} \tag{1.2}$$

Given a Cartesian product, we can define a relation:

Definition 3. *Relation. Let A and B sets. A relation R between A and B is a subset of the Cartesian products $A \times B$.*

The reader might ask, why are we going over these basic definitions? well, all these concepts formally defined are to finally define a function. Remember that functions are those laws and behaviors that neural networks and other algorithms are trying to approximate, that is why it is important to properly define the concept.

---

[2]I am assuming that the reader knows the concept of set and basic set algebra (like subsets, differences, unions, intersections and complements).

Definition 4. *Function. Let A and B sets. A function $f : A \to B$ is a relation R in the Cartesian products $A \times B$ which satisfies that:*

1. *$D_R = A$, in other words, for all $x \in A$ there is a pair $(x, y) \in R$*

2. *Each element $x \in A$ has only one element of B associated with it; in other words $(x, y_1) \in R$ and $(x, y_2) \in R$ imply that $y_1 = y_2$*

The set A is called the domain of the function and B the codomain. And for each $x \in A$, we denote by $f(x)$ the element of B that corresponds to x, i.e. $(x, f(x)) \in R$. And we call $f(x)$ the image[3] of the element x.

Before getting into neural networks and their essence, we need to establish the foundations. Understanding how topological transformations are linked to the spaces they act on (domain and codomain) is a crucial step towards understanding neural networks .

As neural networks are function-approximation algorithms, see [Gurney, 1997] we need to ask ourselves what kind of functions can a neural network approximate? all functions? discontinuous functions? differentiable functions? The answer is that it is in their nature to transform continuously, therefore the functions that they approximate are continuous and even more, to perform backpropagation we require the inverse to be continuous and differentiable; in other words homeomorphisms.

And I would say that not all homeomorphisms; as [Falorsi, 2018] states, it is not always possible to find a suitable homeomorphism

---

[3]Some authors call the image the range of the function

that a neural network can map, especially if data is concentrated at a low-level manifold. Hence, one important question is what kind of homeomorphisms are those that neural networks can approximate.

The above paragraph might sound cryptic: homeomorphism? manifold? What is all of that? Well, we will get into it formally and in an orderly manner in the following chapters.

Even other machine learning algorithms that are not neural networks are based in mathematical concepts encapsulated in the field of probability, statistics, topology and linear algebra.

1.2   Notion of Ergodic Theory

In turn, probability and statistics have their fundamentals on measure theory and ergodic theory, which study the category of measure spaces in which the morphisms (measurable functions, like probability distributions) are actually the transformations that preserve it (preserve measure).[4] However for some authors like Ya Sinai see [Sinai 1994], the main problem of the area is the study of the statistical properties of the objects in groups of non-random movements. We will touch more on the preservation of properties, which in mathematics we call "invariants".

---

[4]A morphism is a transformation, a function or a mapping. The actual etymology of the word means to shape or form (verb or noun), which I find very relevant as morphisms shape spaces through transFORMING elements of those spaces into other elements.

So this book will present a comprehensive study of ergodic theory as a real application of advanced statistics which is the foundation of machine learning algorithms and also of dynamical systems which are fundamental for AI processes and mathematical modeling.

I cannot stress enough that the concepts of measure spaces, morphisms and statistical properties of objects in groups are absolutely crucial in the field of advanced computing,data and decision science and artificial intelligence.

Disclaimer: This is not an easy trip that anyone can do in 3 months; on the contrary, it requires effort, hours of digesting concepts and dedication to study.

This is not a game, and I will not try to make it "fun". It should be fun for you reader if you are genuinely interested in the topic and want to understand the concepts thoroughly.

So let's continue -once again- using some analogies and geometric interpretations of pure analytical theory:

Giving an example of an ergodic system is a bit complex, in fact by studying a very simple system like $z \rightarrow e^{i\sqrt{2\theta}}$ for all z in $\mathbb{C}$ with $\|z\| = 1$ would require a close study of a series of structures in the unit circle that are not obvious in elementary mathematics.

But as an analogy, an ergodic system is that in which a particle moves randomly in a closed space. Imagine a car tire where the molecules of air are trapped inside.

When one moves the car, the molecules start to move. The movement of all molecules (particles) is random. So, the ergodic question is: "if we follow a particle in the tire, after one hour since the car started moving where would the particle be?" and more importantly if we run the car infinitely where would that particle inside the tire end up?

So what we have here in mathematical terms, is a) a closed space (it is important that the space is bounded and does not span to infinity), b) there is movement; i.e. an element in that space, is moving according to c) a mathematical law (a law of movement or physical law) that is always the same and it repeats itself iteratively and is always applied to the element (or particle); in mathematics we represent this *law* as a function[5].

---

[5]As a note aside, the theme of recurrence and repetition has been studied by philosophers and scientists along history from antiquity to modern times, and it is one of the main features in natural phenomena and human activities. Together with Boere, and after discussions with Maxi Zattera I have developed a recurrence theory to explain qualitative change through quantitative iterations.

Neural networks act with iterations (we call them epochs). These iterations are repetitions of a function and at each iteration we transform a function continuously or homeomorphically[6]. So the two fields, ergodic theory and dynamical systems come together very conveniently to lay the foundations of advanced machine learning and artificial intelligence.

1.3   Notion of Spectral Theory

However we need one more element almost always present in all advanced data science algorithms: eigenvectors and eigenvalues. The mathematical field that studies them (eigenvectors and their respective eigenvalues) is called spectral theory.

So, on the one hand we have the study of measurable spaces and ergodic theory i.e. probability and statistics including statistical mechanics; and on the other we have dynamical systems and spectral theory i.e. iterations, functions evolving in time and their operators and special parameters; all this to build solid mathematical models and their connection to "reality". For more example of applications to real-life situations in operations research, and statistical analysis see [Plata, 2008] where we can find the analysis of optimization processes and applied mathematics.

I would like to highlight that spectral theory is not just about finding eigenvectors and eigenvalues, but it tackles the problem of study the geometry of linear transformations $T : V \to V$.

---

[6]Homeomorphically means basically continuously.

In particular in finite dimensions, where it is important to study the problem of finding (if possible) a basis $\gamma = v_1, v_2, ..., v_n$ and constants $\lambda_1, \lambda_2, ...\lambda_n$ such that

$$Tv_j = \lambda v_j \qquad (1.3)$$

Those values λ are called eigenvalues.[7] In this case we say that T is diagonal and the component acts homotethically[8] on all the basic directions and the rest is linearly interpolated. This is not always possible. I will start with the following linear transformation (remember here that matrices are linear transformations) and calculate the eigenvalues and eigenvectors (I will assume the reader has basic notions of linear algebra and can manipulate finite matrices):

$$\begin{bmatrix} 1 & 0 \\ 1 & 1 \end{bmatrix}$$

---

[7]In the following chapters we will go through what a linear transformation is as well as a basis.

[8]Homothetically means that two geometric objects maintain their shapes and ratios when expanding or contracting them. This means that they lie in the same plane and their corresponding sides are parallel; these objects are connected to a point which is commonly known as the vanishing point, but in geometry is called the homothetic center. Girard Desargues (1591-1661) developed the field of projective geometry, which led the formalization of perspective in the work of painters like Della Francesca, or Dürer; arguing that human beings see homothetically or with perspective. More formally, a homothety is a homogeneous dilation of an affine space (affine means preserving parallel lines) determined by a point called center (vanishing point) and a nonzero number λ called its expanding ratio.

The eigen-equation looks like the following where we need to find the values for the transformation:

$$\begin{bmatrix} 1 & 0 \\ 1 & 1 \end{bmatrix} \begin{bmatrix} x \\ y \end{bmatrix} = \lambda \begin{bmatrix} x \\ y \end{bmatrix}$$

Following elementary linear algebra, the system then is solved by calculating the determinant of the matrix as shown:

$$det \begin{bmatrix} 1 - \lambda & 0 \\ 1 & 1 - \lambda \end{bmatrix} =$$

$$(1 - \lambda)^2 = 0 \tag{1.4}$$

The expressions above is called the eigen-polynomial; and solving for λ we get:

$$\lambda = 1 \tag{1.5}$$

And therefore if $Tx = x$ holds, then

$$\begin{bmatrix} 0 & 0 \\ 1 & 0 \end{bmatrix} \begin{bmatrix} x \\ y \end{bmatrix} = \begin{bmatrix} 0 \\ x \end{bmatrix} = \begin{bmatrix} 0 \\ 0 \end{bmatrix}$$

And the above happens if and only if $x = 0$ therefore the space of eigenvectors is:

$$(\begin{bmatrix} 0 \\ y \end{bmatrix} \mid y \in \mathbb{C}) = E_1 \tag{1.6}$$

which has dimension 1 ($dim E_1 = 1$).

One abstraction that must be done in its own right is when $dim V = \infty$, which will not be studied in this book[9].

For many data and decision scientists the concept of eigenvectors and eigenvalues is not alien, so the only thing left to do in this case is to formally introduce the field of spectral theory, which is dedicated to study eigenvectors and eigenvalues.

Finally, it is very important to say that in this book we will study ergodic systems and their eigenvectors and eigenvalues. For this it will be necessary to combine concepts of many areas of mathematics, like group theory, topology, algebraic topology, measure theory and linear algebra. I have done this before with excellent results; an example of a geometric approach to statistical concepts can be found in [Plata, 2006].

This is the truly special characteristic of this work. It is deep in concepts but at the same time it is not specialized in only one field. As a note aside: I was always very frustrated in college when people could only talk about their very specialized field and nothing else and in my many discussions with Alec Boere, I came to the firm belief that the human mind works by association and that is why it is important to get into many fields as possible and look for their interconnections and not being ostracized in one.

---

[9]I am taking for granted that the reader has a minimum knowledge of linear algebra and basic matrix manipulation and operations

1.4   Two Epistemological Theories

The concept of science and the scientific method has changed since Francis Bacon and the empiricist proposed it in the early 1600s. The method was proposed in Bacon's book *Novum Organum*[10] in 1620, and was a response to the classic logical methods proposed by Aristotle in his *Organon*[11] and it is what we know now as the scientific method i.e. observation, hypothesis, experimentation, and theory or conclusion. This method, thanks to our technological advancements is not widely used anymore (specially in decision science). I can understand why the scientific method was such and why it was so successful in those days. But now, with the advanced computational tools and open source shared technology, we can skip a number of these steps.

In general, (and I am not talking about the conscientious data scientists) a data scientist would start with the conclusion, run some data through some algorithms found in a public repository and confirm to a certain degree the conclusion itself. I have rarely seen data scientist hypothesizing anything. For example a typical problem is that one of predictive analytics[12]; say we want to predict when a machine will fail, a

---

[10]"Novum Organum" translates into "New Method".

[11]In the same way, Galileo in his "Dialogo Sopra i Due Massimi Sistemi del Mondo", which translates into "Dialogue Concerning the Two Chief World Systems", opposes to Aristotelian methods and proposes a new way to "make science".

[12]The concept of descriptive and prescriptive methods was originally proposed in the mid 20th century by John Von Neumann [Von Neumann, 1947]; he originally called them "Normative and Descriptive Models" and they were followed in the business world by Clay Christensen as a way to develop theory [Christensen, 2009]. I agree with both authors that descriptive and prescriptive methods are both predictive and I should add that splitting descriptive, from predictive from prescriptive is logically incoherent.

typical data scientist will get the available data and run a regression. If the regression parameters are not acceptable, then they start tweaking the hyper parameters, if that does not work then they change the model and start doing the same cycle.

So no observation, no hypothesis, no experimentation. They go directly to the theory, in the example case "failures have a pattern that can be explained by a regression model" and then work their way backwards.

This new way to make science has some important implications, mainly when we evaluate the ethical value of AI or its responsibility.

Being responsible means that we need to think ethically about our work and proposals, but for that we need a logical or philosophical ground. Hence we need to analyze the problem from a technical perspective and question even the scientific foundations as Henri Bergson [Bergson,1944] did with the theory of relativity.

In this sense, the problem relies in the transition between science and technology: Traditionally AI is focused on the actual meaning of AI and its applications. Responsible AI is then condensed into the technical dimension, i.e. the man-machine relationship as stated by Norbert Wiener, one of the founders of Quantum Mechanics and pioneer of Cybernetics see [Wiener, 1965][13] in which the machine's processes are external to the human, the machine is in general outside the human

---

[13]And this is how Technology works: Analyzing the relationship between humans machines. The scientific method generates the technological environment and its ethical values. That is why mathematics as the science (not the technology) is crucial in the social and human context too.

reach or in some cases it is a mere tool. A follow up is to contemplate all the possible situations where there is a man-machine relationship.

For example, when we are in a car or an elevator; in this case our relationship with the machine changes as we are inside the machine and we become part of the system/process. In a similar way, the data scientist is part of a process more than part of the product.

The main situation is that human understanding in the modern world is driven by our technological ability (this is the main thesis of the work of Gilbert Simondon, see [Simondon, 2017]).

So, if knowledge precedes our ethical perception then only a critical design of a technical development will make the difference from the ethical point of view. This is only one implication of the way modern scientific method acts.

The analogy from the theoretical paragraph above is (using the accordion theory again): "computers are used as nice typewriters" versus "computers changed the way in which we think our writing". The technologically-driven domain dictates our ethical bearings in fact, the technological developments are in themselves ecological as we can see from many examples greatly developed in the work of Gregory Bateson [Bateson, 1972], where he sustains that ecology evolves according to the technical domain; in other words, the technical domain defines the ecological frame. Now the question is "how can we comply with ecological norms if the compliance in itself is the norm?", moreover, how can we build an ecological AI if AI is the ecology itself?

The question deserves more than a simple answer that I will not give here as this is not the purpose of the book, but we can start by "joining the dots".

A fundamental activity in mathematics and AI is connecting the dots. But the dots are not only internal to mathematics or computing, but also to other fields. The true richness is in the real-life human applications; and for this, Jovana Markovic Ioannou, Debolina Guha Majumdar and Lubna Ahbedin are incredibly gifted. People like them deserve all my admiration, always connecting the dots between so many fields and with so many ideas that it is difficult to keep up; but once again, this can only come out of creative and out-of-the-box thinking and not merely being encapsulated in one only field. A great learning from them is how to be versatile and intellectually flexible.[14]

In this same vein, I should recognize the impressive work of Maxi Zattera and Daniel Scain, who not only develop original ideas within the management and technology consulting area, but also in general topics, from cryptography and decoding medieval texts to quantum computing. And this is what I mean by the basis of the creative process and the true connection of dots.

I cannot stress enough how the concepts written in this book are the foundations of highly applicable real-life solutions and should not stay as mere theory. Zattera, Scain, Ahbedin, Guha Majumdar, Markovic Ioannou and Boere are specialists in this with clear applications in

---

[14]It is important to note that to do this one must work in a team and have constant collaboration. It is not just one person but many; Newton's phrase *standing on the shoulders of giants* is of the utmost relevance here. And managing a team to create meaningful outcomes is an art that all the mentioned are experts in.

fields like education, insurance, finance, supply chain, marketing, retail, general production, even life in general and many others. But especially the practical and almost pragmatic application of theories are beautifully explained by Kofi Awuma. I admire his deep understanding of concepts and focused applications.

Finally, two epistemological theories:

1. I got it from my discussions with David Semach: This I call the "accordion theory". An accordion has to be squeezed and then stretched to pump air and make music. In the same way, when working with mathematics, if we only squeeze into the theory or expand into the applications there will be no sound or no knowledge, hence we need to play between theory and examples.

   This is precisely the spirit of the book and you will see examples of it along the book (from notions to formalization and back to notions or from algebra to geometry or from absolute abstract ideas to down-to-earth examples). That is why you will find hard-core theory to understand models' foundations and technological abstractions and almost immediately after an application, example or analogy. This is how I construct mathematical concepts in this book.

   During my after-hours discussions with Martin Weis I learnt that this theory also applies to other situations, for example development and communication[15], thought and action, generalization and distinction, or technology and business; this last

---

[15]And also within communication as one has to state the short term achievements and the long term aims iteratively. This is an art that should be learnt by all data and decision scientists. I found that people with business studies are good at this.

one especially applicable to effectively landing work, without which I truly believe the field of AI would be extremely diminished[16].

2. The theory I like to call "the third iteration recurrent theory (THIRTY)"; this theory is reflected in the underlying topics and concepts of the book; moreover, it is with iterations and repetitions that I explain and connect a wide range of ideas in this book which are contained in themselves in the book, i.e. I explain the concept of *iteration* with iterations!

This approach is fully aligned with other topics of the book, for example recurrence and convergence[17].

Not only does this theory is powerful in mathematics and decision science, but also in many aspects of life. Recurrence and repetition are present in many aspects of business, studies, and personal life and you will see the application of the theory along this book too.

This theory was forged during my discussions for over two years' with Alec Boere, and also the collaboration with Maxi Zattera simply says that we need to iterate at least three times to see convergence: quantitative becomes qualitative[18] at this point just

---

[16]And I think it is like this how mathematics should work, be taught and understood. For more on this same vein of thought, see the *kill math* project by Bret Viktor.

[17]On the philosophical side, the concept of recurrence is widely explore in the work of Nietzsche, Kierkegaard and Deleuze, however, they do not explore the links to convergence. This link might be obvious in the mathematical sense, and it is worth its philosophical exploration in its own right.

[18]There are a number of examples of this even from middle ages with the work of Nicholas of Cusa and its *Coincidentia Oppositorum.*

by iteration and recurrence. Most importantly, this is one way to find the true north in any aspect of life as first iterations can be deceiving and it is only when we recur that we see convergence, or divergence and long term results. Sounds easy but it is not, in fact it is an art more than a science that one has to conquer.

These theories might sound easy, but don't get mistaken, their correct application leads to breakthroughs. In this case, they led to the construction of mathematical concepts like probability leaving aside the classic axiomatic approach. Only by playing with concepts like an accordion with iterations, I reach the mathematical theory in an empirical way.

Therefore, the combination of the theories raise to the epistemological level. This is not only the mere application of concepts, revision of theories or the pure hermeneutical activity (interpretations), but a conceptual proposition to explain mathematics, AI, and both: the intellectual and operational solutions in the real world, making a clear distinction between science, technology and technique from the empirical point of view.

In simple terms, iterations will give us the technique, the function will give us the science and the to-and-fro movements oscillating like a pendulum between abstract and applied (like an accordion) will give us the technology.

Hopefully the reader will see the reflection and application of these theories along the book towards the construction of knowledge. The way I approach knowledge is highly influence by the work of the philosopher Dr. Claudio Canaparo.

I emphasize ideas proposed by Canaparo about "thought and philosophy" in [Canaparo, 2021] and its connections with the concept of geo-epistemology; for more on his seminal philosophical work on the construction of knowledge and geo-epistemology see [Canaparo, 2010].

That is why the connection with many fields of theoretical mathematics and practical applications, are present in a joyful play between analogies, and abstract concepts.

Another idea that hopefully the reader will notice is the topic recurrence and systematic iterations which are present all the way in this work.

Chapter 2

Dynamics: An Introduction to Functions in Time

In existential mathematics, experience takes the form of two basic equations: The degree of slowness is directly proportional to the intensity of memory; the degree of speed is directly proportional to the intensity of forgetting.

Slowness
Milan Kundera

One of the most important geometric objects on which we will base this study is the torus. The torus is like a doughnut with a hole in the middle, which thanks to these homeomorphic transformations (we briefly talked about them in the previous section and introduction) can be transformed into the two-dimensional space.

That is why the torus is so important. In other words, the torus is the two-dimensional plane!

The homeomorphisms and flows on the Torus are very important from many points of view. These are very special classes of dynamical systems, as we will see in the next section. However we need to formalize certain concepts to understand them thoroughly and lay the foundations to launch our knowledge further; so let us start with the following definitions [1]:

2.1   Discrete Dynamical Systems

Following the spirit of the accordion theory, I will start with a general notion of dynamical systems to follow with the formalization.

In this way we will be able to grasp the subtleties of the technique and the scientific concepts behind them.

These systems are functions that repeat as iterations over themselves. The previous result is the argument of the next repetition and so on. Do not over-complicate this from the start, the notion of a dynamical system is an iteration of a function.

The many results of the function form an orbit and that is what in general we refer as a dynamical system.

---

[1]I will also assume a basic understanding of the mathematical notation, number sets and basic concepts like composite functions, identity elements, etc.

Continuous dynamical systems are differential equations. In fact if you think of the basic radioactive alpha-decay equation[2], we have:

$$\frac{dM}{dt} = -\lambda M \qquad (2.1)$$

where M is the mass of radioactive material and λ is the constant of proportionality.

This equation is describing how things change in in time, in other words, functions iterating in time.

The solution of this equation is:

$$M(t) = M_0^{-\lambda t} \qquad (2.2)$$

Where M_0 is the initial condition. Well, the initial condition is precisely the first value of the dynamical system. If we break the equation above in its iterations we obtain that:

$$M_1 = M_0 \cdot -\lambda \qquad (2.3)$$

and

$$M_2 = M_1 \cdot -\lambda = M_0 \cdot -\lambda^2 \qquad (2.4)$$

---

[2]Proposed originally by F.Soddy and E.Rutherford

and in general

$$M_{n+1} = M_n \cdot -\lambda \qquad (2.5)$$

In other words, it is an iterative function. In this book I will only work with discrete dynamical systems and leave the continuous case just as a reference.

Before tackling the formal theory, I would like to make a distinction between datum, data, information and knowledge. Datum is the atom of abstractions; it is the minimal unit with which we abstract the world so we can make sense of it[3]. Abstractions can be representations or even mental images out of real-life objects, phenomena, or processes. Traditionally a datum has been always numerical and even when images can be translated into matrices of numbers, we can consider now that data can be also words, images, films, numbers and relationships (like graphs or networks).

We produce data to explain and reason our universe. For example, the Earth going around the sun one time is translated into 365 days. Abstractions like this allow us to study the behavior of our universe.

The plural of datum is data. So data is the aggregation of datum. But data on its own does not generate any knowledge. Information has to

---

[3]There are three main level of abstractions in philosophy: a) the physical one, which is all of what we perceive with our senses. Perception is the first level of abstraction, all what we see or hear or touch, etc. are examples of it b) the second level is the mathematical one, examples of this level of abstraction are the concept of infinity, or operational syntax and c) the third level of abstraction is the metaphysical, examples are the concept of soul or freedom.

be drawn out of data before we can generate knowledge. Examples of information are organized patterns of data, for example data points in time (like a time series), or orbits of planets in a chart.

Out of them we generate knowledge. Knowledge is a construction of narrative out of the information. As you can see, there is a logical chain and layers, which come from the atomic particles (datum) to an entire narrative which constitutes knowledge. For more see [Plata, 2007].

Now I will start with the formal theory:

Definition 5. *Discrete Dynamical System. Let X be a set $\phi : \mathbb{Z} \times X \to X$, if $T : X \to X$ is an arbitrary function and $\phi(n, x) = T^n(x)$ then T is a discrete dynamical system.*

As an example of a dynamical system take a real number and calculate the cosine of that number, then calculate the cosine of the results y subsequently calculate the cosine of the following results. We will obtain the following:

$$\underbrace{cos(cos(cos(cos(cos(...cos(x))))))}_{\text{n times}} \qquad (2.6)$$

In this dynamical system one can forecast that the first value, after calculating the first cosine, will fall in the interval $[-1, 1]$; as a matter of fact, the second one too as cosine is a bounded function between -1 and 1.

One of the most important things when we study dynamical systems is how the iterations of the function (or the law of movement) map the values of the function, and in general the asymptotic behavior of a system would be very relevant. For more on dynamical systems see the seminal text [Devaney, 2003].

In order to outline these iterations and their values when acting on elements, one needs to study the *orbits* of the system. The orbits are precisely the values where the function iterates. The definition will be divided in forward orbits and backward orbits and is the following:

Definition 6. *Orbit. The forward orbit of x is the set of all points $x, T(x), T^2(x), \ldots$ and we denote it as $O^+(x)$. In other words, $T^n(x) = \underbrace{T \circ T \circ T \circ T \circ T \circ T}_{n\ times}$. For the case of bijections, the complete orbit is the set of all points of the form $T^n(x)$ with $n \in \mathbb{Z}$. That is, one can define the backward orbit $O^{-(x)} = x, T^{-1}(x), T^{-2}(x), \ldots$*

circle = S^1

orbit on the circle

The simplest orbits come from the following three definitions:

Definition 7. *Fixed Point. The point x is a fixed point of T if $T(x) = x$.*

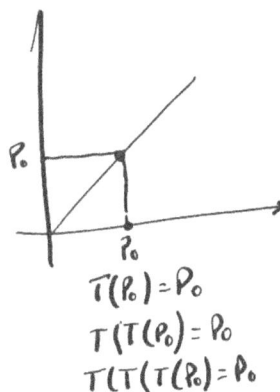

$$T(P_0) = P_0$$
$$T(T(P_0)) = P_0$$
$$T(T(T(P_0)) = P_0$$

For example if $T : \mathbb{R} \rightarrow \mathbb{R}$ is the identity function, all the points are fixed since $\forall x \in \mathbb{R}, T(x) = x$ holds.

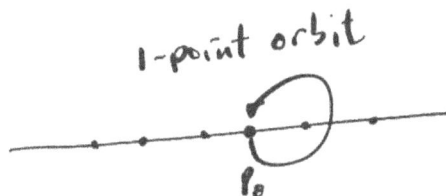

1-point orbit

Definition 8. *Periodic Point. The point x is a periodic point of T if $T^n(x) = x$, for a given $n \in \mathbb{Z}^+$*

In other words, if x is periodic, then $\exists n \in \mathbb{Z}^+ \mid T^n(x) = x$ and the lowest positive integer for which this happens is called the period of the orbit.

Sometimes the periodic points come after a number of iterations, that is why it is important to investigate the asymptotic behavior of the systems and their orbits. For example a particle can travel for some time ($t = 1, 2, ...n$) not showing a specific orbit, however after n iterations there can be a cycle or three points or something more specific.

If n is the period of the orbit of T, then $x, T(x), T^2(x), ..., T^{n-1}(x)$ are different points.

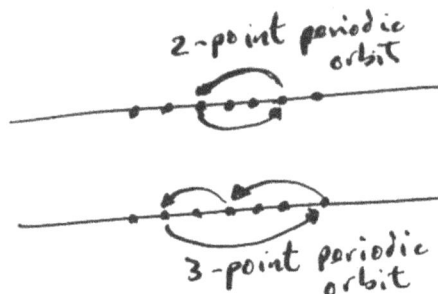

Definition 9. *Final Periodic Point. A point x is a final periodic point of period n of T if x is not periodic, but $\exists m > 0$ such that $T^{n+i}(x) = T^i(x), \forall i \geq m$.*

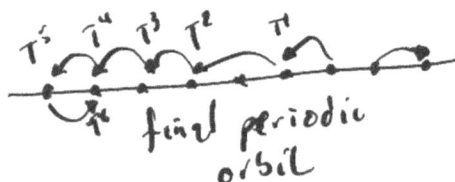

A final periodic point would be a convergence point. Particles in a dynamical system can behave in many ways before convergence, so when studying them we are interested in identifying changes in behavior like bifurcations, and in general, their asymptotic behavior; this last one happens when we iterate the system largely or more precisely when the number of iterations n approaches ∞.

One objective of studying dynamical systems is understanding the nature of the orbits and find out if these are periodic, finally periodic, etc. Doing this is difficult and depends on the nature of the system, for example if one wants to find the periodic points of period n in a dynamical system, whose function is a quadratic one, one would have to solve the equation $T^n(x) = x$ that is a polynomial equation degree 2^n.

In some occasions (when one can graph the function), finding fixed points or periodic ones in a dynamical system can be done via graphical analysis that we will describe in the next example.

Although this technique does not give us the exact values of the points, helps us understand qualitatively the nature of the system.

The graph the function $cos(x)$ -like in the example given above- where the dynamical system is $T(x) = cos(x)$ can give us a great amount of information about the first iteration, but not the subsequent.

It is not obvious how to find in which point the system will land after iterating the function n times; in other words, finding which points are fixed or periodical on $T^n(x) = x$.

2.2   Asymptotic Behavior

In order to identify the points, one needs to graph the function of the dynamical system together with the diagonal line at 45 degrees $\mathscr{D} = \{(x, x) \mid x \in \mathbb{R}\}$; in other words, the line on $\frac{\pi}{4}$ or the identity function $f : \mathbb{R} \to \mathbb{R}$ such that $f(x) = x$.

To illustrate this, we will continue with the example of $cos(x)$ and we will outline the steps to find the orbits of the dynamical system generated by the function.

1. Suppose that we start at any point p of the domain of the function (let's remember that the natural domain of the function $cos(x)$ is $\mathbb{R}$). At the first iteration we will end up in the interval between zero and one $[0, 1]$

2. Remember that we will calculate the cosine of the number found in the previous iteration and so on, then we can start finding the intersection with the function; this is found by following a vertical line until hit the graph of the function at the point $(p, T(p))$

3. From the point $(p, T(p))$ one has to draw a horizontal line until we intersect $\mathscr{D}$ (remember this is the diagonal at 45 degrees), then we will be exactly at the point $(T(p), T(p))$

4. After this, we will follow a vertical line until we intersect once again the function; then we will be at the point $(T(p), T(T(p)))$, hence finding the points of the second iteration

5. Then we will repeat the process following steps 2 and 3 iteratively. To illustrate the next point specifically, we will construct a horizontal line from the last point found until hitting $\mathscr{D}$, in this case we will be at $(T(T(p)), T(T(p)))$

In this way we can find where the nth iteration of the function intersects the identity and hence if the system diverges or converges towards a point, or any other behavior.

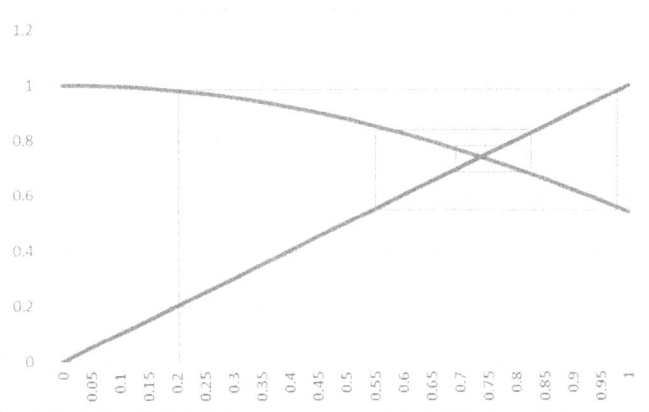

Another example is the following:

Suppose that we want to analyze the system $T(x) = x^2 - 0.7$ and we start at the point $x_0 = 0.75$. Following the process described above, the system will approach to a value between one and zero the more we iterate the function. This can be seen in the following graph. This is called the graphic analysis of the dynamical system T.

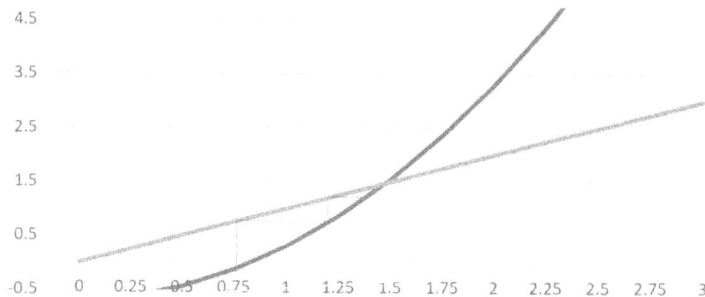

However if the initial point is $x_0 = 1.75$ then the process diverges to ∞.

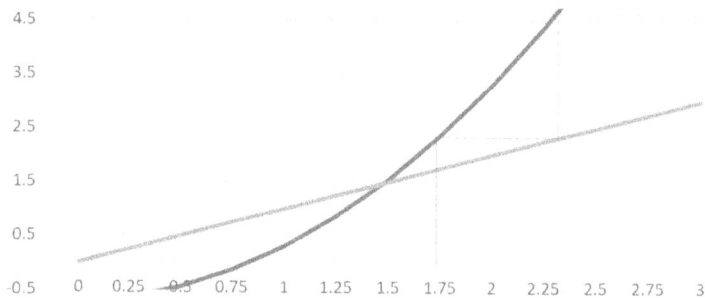

To illustrate a system with a periodic orbit of period 2 we can have a dynamical system $T(x) = x^2 - 1$; the reader will note that $T(0) = -1$ and $T(1) = 0$.

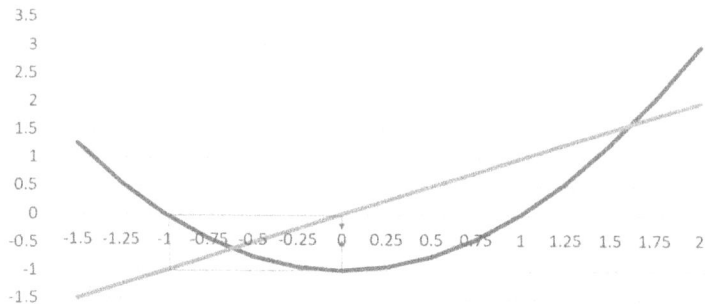

Therefore it is also relevant the initial conditions (or starting points).

In this book we (you, the reader and me) will study the discrete dynamical systems and I will leave out the continuous ones. I will only remark that a continuous dynamical system is called a differential equation.

Now we can start seeing the links between different fields of mathematics. Dynamical systems are related to calculus too. And this is how we start to connect the dots.

A very good example of how dynamics work and is related to calculus can be found in the outstanding work of [Bisewski, 2022] where the author treats the case of fractional Brownian motion.

Chapter 3

Groups: An Introduction to Algebraic Structures

> Mathematics is the queen of the sciences and number theory is the queen of mathematics.
>
> *Waltershausen's Biography*
> *Carl Friedrich Gauß*

In general algebra studies structures. All structures? well, meaningful structures; structures that lead to something important, like the space of all continuous functions, the mappings that can be inverted or commute, etc. Algebra is the grammar of mathematics[1].

---

[1] I will assume that the reader is also familiar with basic notions of set theory and operations i.e. addition and multiplication. I will also assume that the reader is familiar with vector operations, like multiplication by scalar and cross product.

Grammar studies the structure of language and its parts, like adjectives, nouns, verbs, and so on, and how they play their part in sentences, for example what elements should be placed in the direct object, or how to construct adverbial phrases, etc.; in the same way, algebra studies mathematical structures and its elements[2] and how they interact with each other.

I see syntactic analysis in general grammar as equivalent to syntactic analysis in mathematics. Both study syntax in pretty much the same way. Algebra in middle school teaches us how to manipulate the syntax of mathematical forms like equations. So hopefully the reader will spot the similarities now.

You might find irrelevant to learn the prepositions, or the correct use of possessive pronouns, while studying English grammar, but in the end one understands that to have full domain of the language one needs to understand its structure (and that happens with any language, especially when learning a second or third one).

In the same way, the reader needs to understand what an inverse or neutral element is and in the end one needs to understand that too have full domain of mathematics one needs to understand its structure[3].

---

[2] Abstract algebra links fundamental concepts in many fields of mathematics, like geometry , number theory and basic algebra. So the study of those structures is crucial to understand solutions in all of those areas of mathematics.

[3] This would be a necessary but not sufficient condition.

3.1   Structures

So let us start by learning the mathematical structures, in our case algebraic structures, that are important and meaningful and lay the foundation of many theories.

One of the most important and fundamental structures is the so-called group.

Normally in algebra we operate elements two at a time; in other words, we have binary operations all the time. For example there is no such thing as 2+3+7, you see it is not a binary operation, it has three elements! what we do in mathematics is first operate 2+3, which equals 5 and then 5+7 to get the final result. It is always two at a time.

Obviously in order to do that we need to make sure that grouping is allowed and leads to the "correct" result; in other words, we need to make sure that (2+3)+7 is allowed.

There are a number of other properties that algebraic structures have, so we will start by defining an algebra; not algebra the subject that you study in middle school, but an *algebra*; which is a mathematical object or structure.

Firstly we need to know that there are structures like monoids, rings, fields and algebras . The simplest structure would be a monoid in which we have a set of elements and a binary operation, that is it! no more, just elements and an operation, no inverse, no association, nothing, only the binary operation.

This is the simplest of all structures. Mathematical methods generally start from simple concepts and grow on complexity and extension; so I will follow the same methodology to explain structures.

The process to build more complex and meaningful structures consists of adding more properties to the more simple ones so we make them richer and more powerful to *do more things*.

Now I will start constructing structures from the simplest to more complex, so the reader can see the mathematical process, and most importantly how the *theory of the recurrent iterations*[4] works.

I will iterate by adding properties to the previous structure. In this way I will build full mathematical structures through the process of dynamical systems more than being axiomatic and merely defining them without context. So let us start with the formal process and method. The formal definition of a monoid is:

Definition 10. *Monoid. A monoid is a set A equipped by a binary operation $\oplus$ (it works like the one described above (2+3)+7) and for which any two elements of A the result of the operation is also in A. This is called that the set is closed under an associative binary operation. There is also a neutral element I in A such that for all a in A, the operation $I \oplus a = a \oplus I = a$. In other words, a neutral element when operated does nothing.*

---

[4]More on the theory is explained in chapter one where Alec Boere and myself discussed it.

And in the spirit of writing mathematics we also need to define a binary operation. Can you see how this works? We need to define basically everything so we can proceed with further knowledge and avoid ambiguities.

Definition 11. *Binary Operation. Given a set A, let $A^2 = \{(a, b) \mid a, b \in A\}$. A binary operation is a function from A^2 to A.*

When we impose more structure to the set and we define more given operations in the set we advance in meaning and are able to do more things with the set.

I think this is a good place to utter one definition of mathematics. The reader might find that some definitions of mathematics refer to the science concerned with numbers, but how about geometric shapes? Well, we can add that too, but how about functions and change? Well, that too, and how about vector spaces? Well, that too, and how about applications (like expressing phenomena in physics)? Well that too! But if you ask someone who studies algebraic topology, functional analysis or differential geometry, they would not agree with any of the previous definitions. Moreover, for someone with computer science background would argue that formal logic is also part of mathematics. So what is mathematics? As you can see mathematics (as any other complex subject would be very hard to define. But I will try to simplify the definition in this context. "Mathematics is the study of meaningful structures".

So let's continue with our study of structures: An important definition is the one of a "field". A field is a "ring" -which is another structure in the algebraic ladder- but with a bit more added properties.

Definition 12. *Field.*[5] *Let F be a set, with two binary operations $+$ and $\cdot$. F is a field if the following conditions hold:*

1. *If $x \in F$ and $y \in F$, then their sum $x + y$ is in F. This property is called additive closure*

2. *$x + y = y + x$ for all $x, y \in F$. This property is called additive commutativity*

3. *$(x + y) + z = x + (y + z)$ for all $x, y, z \in F$. This property is called additive associativity*

4. *F contains a neutral element 0 such that $0 + x = x$ for every $x \in F$. This property is called additive identity and 0 is called the additive identity*

5. *To every $x \in F$ there is a corresponding element $x \in F$ such that $x + (-x) = 0$. This property is called the existence of an additive inverse. In other words, an element plus its inverse is the identity element*

6. *If $x \in F$ and $y \in F$, then their product $x \cdot y$ is also in F. This property is called multiplicative closure*

7. *$x \cdot y = y \cdot x$ for all $x, y \in F$. This is multiplicative commutativity*

---

[5]This is a classic definition that can be studied in a proper analysis course following the classic text by Walter Rudin, see [Rudin, 1974].

8. $(x \cdot y) \cdot z = x \cdot (y \cdot z)$ *for all* $x, y, z \in F$. *This property is called multiplicative associativity*

9. F *contains an element* $1 \neq 0$ *such that* $1 \cdot x = x$ *for all* $x \in F$. *This property is called the existence of a multiplicative identity. 1 is called the neutral element for the multiplication*

10. *If* $x \in F$ *and* $x \neq 0$ *then there exists an element* $\frac{1}{x} \in F$ *such that* $x \cdot \left(\frac{1}{x}\right) = 1$. *This property is called the existence of a multiplicative inverse. and* $\frac{1}{x}$ *is called the inverse element*

11. $x(y + z) = x \cdot y + x \cdot z$ *for all* $x, y, z \in F$. *This property binds the two binary operations* $+$ *and* $\cdot$ *and it is called Distributive law*

Subtraction is a special case of addition and division is a special case of multiplication.

A field is a set with addition, subtraction, multiplication and division, and the set of real numbers, denoted by $\mathbb{R}$ is a field.

Every field is an algebra because every field is a (one-dimensional) vector space, but not every algebra is a field.

Just to give the reader an idea, of how structures in mathematics work, a "ring" which is a set with two binary operations that does not have multiplicative inverse comes from a *rig* or *semi-ring* with added properties.

If we keep adding more properties to the ring, then we will get a field; a field, is a commutative ring where every nonzero element is invertible.

So the next structure is given by the previous one with added properties under an iterative process in the same manner as dynamical systems.

An algebra is a ring with added structure with respect to scalars (scalars are just plain numbers). Remember that the multiplication by a scalar is a well-known operation with vectors[6]. The reader might notice now that by adding properties to the previous and "simpler" structure we obtain a more solid, powerful and useful one.

A group is a set equipped with an operation that combines any two elements of the set to produce a third element of the set, in such a way that the operation is associative, an identity element exists and every element has an inverse.

Now, the reader might have noticed that there is another set with some structure called "group". Why do we study groups? Well, because groups are the spaces where functions with their inverse "live". As we will see in the next section (topology), inverse functions will help us to define continuity. Groups will also be very useful when studying transformations on complex structures, where computational mappings act, for example the unit circle.

---

[6]I will not define specific sets and structures as this is not the purpose of the book, but I have to mention their existence and hierarchy. It will be important to know this as the basis of probability theory is the so called sigma-algebra or σ-algebra; so the reader should be acquainted with the basic concepts of abstract algebra.

3.2   Formalization of Groups

Groups are fundamental in the study of algebra and support theories in many branches of mathematics. For example, not only is the unit circle the set of points in $\mathbb{R}^2$ such that $x^2 + y^2 = 1$, but also has a more robust algebraic structure that we will illustrate later; however to do this we must base all on "groups" and other structures.

This might sound very abstract, but in mathematics we need to formalize concepts to keep growing knowledge and communicate in a universal way. So let's continue with the following formal definitions:

Definition 13. *Group. A group is a set G subject to:*

i) An associative binary operation $*$ with: ii) An identity element i.e. an $e \in G$ such that $g * e - e * g = g \forall g \in G$ and where iii) All element is invertible i.e. $\forall g \in G, \exists g^{-1} \in G$ such that $g * g^{-1} = g^{(-1)*g=e}$. In general the symbol $*$ is omitted and the operation is normally denoted as gh instead of $g * h$. iv) It is said that a group G in commutative or Abelian if $*$ is commutative for all $g, h \in G$. In other words, $gh = hg$.

In our case , which is the study of homeomorphisms of ergodic transformations, only Abelian groups will be used, which are the commutative groups [7].

This structure will allow us to calculate inverses and commute elements between them as we will see in the following chapters[8]

---

[7]Abelian = commutative

[8]As the reader might have noticed by now, we are outlining meaningful structures,

There are many examples of groups. One of the easiest is the integers $\mathbb{Z}$ with the sum $+$ as a binary operation. In geometry, the rotation of a triangle is also very simple but illustrative, but as a more relevant example to this work we have the circle S^1 with respect to the multiplication in $\mathbb{C}$.

Another example of group is when we calculate with modular arithmetic (also called clock arithmetic). The main concept when studying modular arithmetic is congruence.

Definition 14. *Congruence. We say that two integers a and b are congruent modulo m if m divides the difference $a - b$.[9] in symbols:*

$$a \equiv b(mod(m)) \ \ if \ and \ only \ if \ m \mid a - b$$

If $\mathbb{R}(mod(2\pi))$ is the set of all equivalence classes $[\theta]$ where $\theta_1 \sim \theta_2(mod(2\pi))$ if and only if $\theta_1 - \theta_2 \in 2\pi\mathbb{Z}$, then it is easy to see that the following diagram commutes:

$$
\begin{array}{ccc}
 & \mathbb{R} & \\
\swarrow & & \searrow \\
\pi & \longrightarrow & S^1
\end{array}
$$

---

i.e. structures where we can launch knowledge and functionality. This refers to the previous definition of mathematics. It is not the study of all structures, but the meaningful ones.

[9]In computer science, the modulo operation returns the remainder of a division between two numbers. This is also called the modulus.

There is a property in group theory that establishes that the Cartesian product of groups is also a group. So, given this, we can conclude that the torus $S^1 \times S^1$ is a group with the following operation

$$(z_1, w_1) * (z_2, w_2) = (z_1 z_2, w_1 w_2) \tag{3.1}$$

for all $z_1, z_2, w_1, w_2 \in S^1$

Expanding from the previous chapter, the Cartesian product of sets A and B, denoted by $A \times B$ is a concept in set theory that generates ordered pairs out of the two sets.

Remember that based on the Cartesian product, we defined first a relation then a function and then three main types of functions. So let's recap:

Definition 15. *Relation. Let A and B be sets, a relation R is a subset of the Cartesian product of $A \times B$.*

Definition 16. *Function. Let A and B be sets, a function $f : A \to B$ is a relation R of the Cartesian product of $A \times B$ that satisfies:*

i) The domain of the relation D_R is A; i.e. $D_R = A$. In other words, for every $x \in A, \exists (x, y) \in R$

ii) Each element $x \in A$ is associated only with one of B. In other words, if $(x, y_1) \in R$ and $(x, y_2) \in R$ implies that $y_1 = y_2$

The set A is called the domain of the function and B the codomain. And for each element $x \in A$ we denote $f(x)$ the element in B that corresponds to x; i.e. $(x, f(x) \in R$. We call $f(x)$ the image of x. In classic texts, the codomain is called range, however in modern texts, range is usually meant to be the image. I personally prefer the differentiation between terms so I will avoid the use of the word "range" and I will use the concept of "image".

Definition 17. *Image of a Function. The image Img of a function* $f : A \to B$ *is*

$$Img = \{b \in B \mid \exists a \in A \ni f(a) = b\} \tag{3.2}$$

Now let us get into the main types of functions:

Definition 18. *Injective Function. A function* $f : A \to B$ *is injective if for all pairs* $a_1, a_2 \in A$ *with* $a_1 \neq a_2$ *then* $f(a_1) \neq f(a_2)$.

An equivalent statement is: $f(a_1) = f(a_2) \implies a_1 = a_2$.

Definition 19. *Surjective Function. A function* $f : A \to B$ *is surjective if* $Img_f = B$. *In other words, that for every* $b \in B$ *there is* $a \in A$ *such that* $f(a) = b$.

Definition 20. *Bijective Function. A function* $f : A \to B$ *is bijective if it is both, injective and surjective.*

Finally let us define the composition as this will be useful for further concepts and it is widely used in data modeling.

Definition 21. *Composite Function. Let $f : A \to B$ and $g : B \to C$. We define the composition of f and g, denoted by $f \circ g$ as the function $f \circ g : A \to C$ given by*

$$f \circ g(x) = g(f(x)) \forall x \in A \qquad (3.3)$$

Now let us continue with the abstract algebra definitions:

Definition 22. *Subgroup. If $\emptyset \neq H \subset G$ and H is closed under $*$ and the inversion of elements, then we say that H is a subgroup of G.*

Definition 23. *Permutation Group. Let X be a set. The set of bijections of $X \to X$ is a group under the composition of functions. This group is called the permutation group of X and is denoted by $S(X)$.*

Definition 24. *Homomorphism. Let G and H be groups. A function $\phi : G \to S$ is called a homomorphism if $\phi(gh) = \phi(g)\phi(h), \forall g, h \in G$.*

Please note that a homomorphism is not a homeomorphism. Homomorphisms are structure preserving transformations, while homeomorphisms are bijective continuous functions in topological spaces. The best way to illustrate what a homomorphism is, well, is in a Cayley table where elements transformed under the homomorphism would be in the same position as in the original table under a given operation. In other words, they would not move or preserve the same structure.

The typical example of a homomorphism is $\phi(x) = x^2$

Times	0	1	2	3	4
0	0	0	0	0	0
1	0	1	2	3	4
2	0	2	4	6	8
3	0	3	6	9	12
4	0	4	8	12	16

Now, applying the transformation to each number, we get the squares of the numbers. So the multiplication table looks like this:

Times	0	1	4	9	16
0	0	0	0	0	0
1	0	1	4	9	16
4	0	4	16	36	64
9	0	9	36	81	144
16	0	16	64	144	256

So $3 \cdot 2 = 6$ which is in the coordinate of (3,4) i.e. third row and fourth column of the first table.

Now $3^2 \cdot 2^2 = 9 \cdot 4 = 36$ and 36 which is 6^2 is on the third row and fourth column of the table. If you try any other multiplication you will see that all the elements' places in the tables coincide. Try $3 \cdot 4 = 12$ and see where in the second table is the value for 12^2.

So $\phi(x) = x^2$ is a homomorphism of integers under the usual multiplication.

Definition 25. *Action. If $S = S(X)$ and $\phi : G \to S$ is a homomorphism, then it is said that G acts on X through ϕ and ϕ is an action of G on X.*

It is usual to denote the action by $\hat{\phi} : G \times X \rightarrow X$ such that $\hat{\phi}(g, x) = \phi(g)(x)$ or $(g, x) \mapsto \phi(g)(x)$.

When we say that a something acts on a group it is not a figure of speech, but a very precise thing that is happening (according to the definition above).

Actions on groups are typically dynamical systems or series that iterate, for example $a_{n+1} = \sqrt{a_n} + 3$, so the next element in the series is defined by the previous one. One can observe that the numbers of the series are in the real numbers but the iteration number is in the natural numbers. So we can say that the integers are acting on the real numbers through the correspondence rule.

Definition 26. *Right Lateral Class. Let H be a subgroup of G. The right lateral class of $g \in G$ under H is the set:*

$$gH = \{gh \mid h \in H\} \tag{3.4}$$

All these lateral classes form a partition in G that we will denote by G/H.

Definition 27. *Kernel. The nucleus, core or Kernel of ϕ denoted by $Ker(\phi)$ is the subgroup:*

$$Ker(\phi) = \{g \in G \mid \phi(g) = e\} \tag{3.5}$$

where e is the neutral or identity element.

3.3 Groups and Matrices

Examples of groups are the ones called matrices. If $M(n, \mathbb{R})$ is the set of all order n square matrices with real coefficients, then the following are groups under the multiplication of matrices:

$$G(n, \mathbb{R}) = \{A \in M(n, \mathbb{R}) \mid det A \neq 0\} \tag{3.6}$$

or the group of invertible matrices.

$$O(n, \mathbb{R}) = \{A \in GI(n, \mathbb{R}) \mid AA^t = Id\} \tag{3.7}$$

or the group of orthogonal matrices consisting of all endomorphisms that preserve the Euclidean norm.

$$SL(n, \mathbb{R}) = \{A \in GI(n, \mathbb{R}) \mid det A = 1\} \tag{3.8}$$

or the group of volume and orientation preserving linear transformations.

This group is also called the rotation group, and if we add more structure and require the group to have the identity then it becomes the special orthogonal group $SO(n, \mathbb{R})$, formally defined as

$$SO(n, \mathbb{R}) = O(n, \mathbb{R}) \cap SL(n, \mathbb{R}) \tag{3.9}$$

or the group of orthogonal matrices with determinant equal to 1. For example, the group $O(2, \mathbb{R})$ has a group action on the plane that is a rotation.

Please observe that the matrix

$$\begin{bmatrix} 1 & 0 \\ 1 & 1 \end{bmatrix}$$

of the previous example is in $SI(2, \mathbb{R})$.

And the correspondence

$$\begin{bmatrix} cos(\theta) & -sin(\theta) \\ sin(\theta) & cos(\theta) \end{bmatrix}$$

defines an isomorphism[10] of groups between S^1 and $SO(2, \mathbb{R})$.

After this section it would be easy to see that $GI(n, \mathbb{R})$ or any of its subgroups act on $\mathbb{R}^n$ with the rule $(A, \bar{x}) \mapsto A\bar{x}$.

As a final note in the chapter, it is important to highlight that dynamical systems in a strict sense are actions of groups on sets under given functions.

When we say that the natural numbers $\mathbb{N}$ act on the real numbers $\mathbb{R}$ through a function $f : \mathbb{R} \to \mathbb{R}$, we mean that we iterate over and over the same function or we create a dynamical system. The action is basically the iterations of the real-valued function.

---

[10]I will explain more on isomorphism in the following chapters.

Chapter 4

Topology: An Introduction to Continuity

Geometry is the right foundation of all painting... but we must resign ourselves to the opinion and judgment of men.

The Four Books of Measurement
Albrecht Dürer

I would like to preface this section by reiterating that the book does not concentrate on a specific area of mathematics and the concepts in this book are linked not to one but to many disciplines within mathematics. Other texts are specialized in only one topic or area and tend to deepen that knowledge solely, but not this one. That is why the reader will notice elements of abstract algebra, topology, and dynamical systems so far, but this is how we will connect the dots.

I believe that when one specializes more and more, mainly in the academic world, one tends to narrow their knowledge and "know more about less" instead of connecting the dots with wider concepts. After my discussions with Guha Majumdar and Markovic Ioannou and Awuma[1] I concluded that the human mind works by association, thus if we concentrate on one and only one topic, associations would be nullified.

This is why, we need a diversified knowledge of independent disciplines that in their own right deserve being studied, in our case, group theory, dynamical systems, measure theory, topology, etc. but will be linked logically to shed light into more technical knowledge and will allow the reader to create.

Therefore in these preliminary sections I will cover different themes and concepts that we will use in ergodic & spectral theories and computer science.

At university I attended a conference titled "Eleven Definitions of Topology" which led me to the realization of the complexity of the subject. Indeed, trying to define or elaborate on topology, would be almost impossible without the corresponding years of study of the subject at a higher level. So I will try to explain this in a rather direct way:

One of the most important fields in mathematics (and many researchers would agree with this) is differential equations. Differential equations are related to dynamical systems (as explained in the previous section), and a number of applications to physics, biology and moreover, data

---

[1]See acknowledgements, page i.

science is closely related to them, for example the interesting research by Raissil, et. al. (see Raissil, 2018) linking differential equations with neural networks.

The main tool to solve differential equations (algebraically) is calculus; you know, the subject with the derivatives and integrals, and one of the main premises and we would say the basis for calculus is continuous functions.

Derivatives are possible only because we are dealing with continuous functions. Without them there is no derivative and hence no calculus.

And what does this have to do with topology? Well, topology is where the continuous functions "live". The study of topology is of the utmost importance because it studies continuous functions and their behavior. Now the reader can realize that topology is crucial in mathematics.

Also topological transformations are special because they lead us to important transformations like isomorphisms. Isomorphisms are important because they preserve structure and measure from one space to another[2].

When Einstein was studying relativity theory, he was looking for isomorphisms, or invariant from one space to another.

---

[2]In differential topology there are three types of transformations of mappings: a) area-preserving mappings, b) conformal mappings (they preserve angles) and c) isomorphisms (they preserve area and angles).

Do you see now how important topology and homeomorphisms are? Homeomorphisms will help us establish topological equivalence, without which backpropagation in neural networks would not be possible.

4.1   Formalization of Topology

I guess the reader is realizing a pattern here: we are studying structures and spaces where mathematical objects "live" and also things that preserve structure; in other words, invariants. This note is very important because the study of invariants is incredibly important not only in mathematics, but also in science.

So let's start with the tough abstract concepts:

Definition 28. *Topology. Let X be a set. A topology on X is a collection τ of subsets of X, called "open" (I will define what an open set means in the following definitions), that hold the following properties:*

    1. If A is any set, and $U_\alpha \in \tau$, $\forall a \in A$, then $\cup_{\alpha \in A} U_\alpha \in \tau$

    2. If $U_\alpha \in \tau$ for each α in some finite set F, then $\cap_{\alpha \in F} U_\alpha \in \tau$; and

    3. $X \in \tau$ and $\emptyset \in \tau$

Why arbitrary unions and finite intersections? Well, to avoid exceptions, like the set $\cap_{n=1}^{\infty} \{-\frac{1}{n}, \frac{1}{n}\}$ which is a countably infinite set of open intervals whose intersection is not open.

We will get into the proper definition of open sets and refer to interesting examples like the Cantor set to understand better these concepts.

But first a note: in the same way as algebra, do not confuse topology as a mathematical discipline with *a topology*. As I explained in the introduction, this is how mathematicians deal with duality and ambiguity. It is a great training when consulting in business and one has to manage ambiguity.

Definition 29. *Topological Space. A topological space is a pair (X, τ) where τ is a topology on X.*

Definition 30. *Continuous Function. Let $f : X \to Y$, it is said that f is continuous if $\forall V \in \tau_Y$ we have that $f^{-1}(V) \in \tau_X$*

Note that V is a set not a single value or point. So we are talking here of entire sets mapping into other sets. This is one of the magical features of topology. This will be very useful in computing as the computation can be done in entire sets and not just point by point.

Ok, stop for a second! Can you see now the connection with the previous section where we defined inverse functions and their habitat? Now we need inverse functions to define continuity!

Another thing to note - and this is for those initiated into the deep knowledge of mathematics - is that continuity traditionally has to deal with limits and concepts defined by epsilons and deltas which are part of the classic definitions of limits, continuity and infinitesimal calculus.

So just for the sake of it, let's remember that definition from more basic mathematics:

Definition 31. *Continuous Function at a Point. We say that $f : \mathbb{R} \to \mathbb{R}$ is continuous at $a \in \mathbb{R}$ if and only if $\forall\ \epsilon > 0, \exists \delta > 0$ such that if $\mid x - a \mid < \delta \implies \mid f(x) - f(a) \mid < \epsilon$*

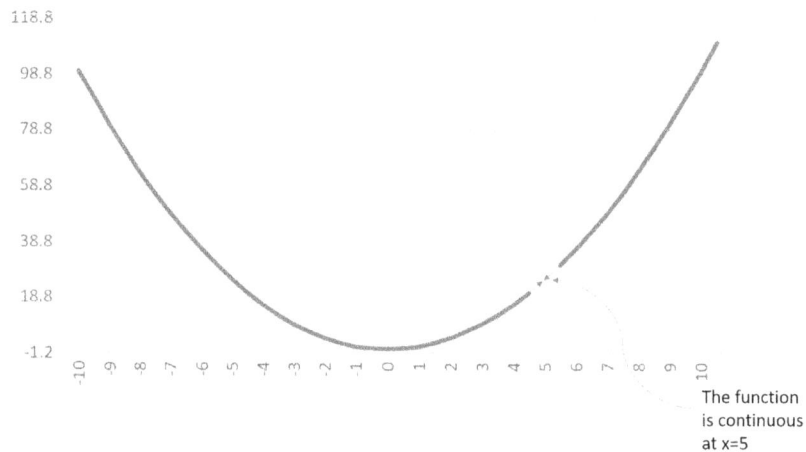

The function is continuous at x=5

As you can see from the image the graph "follows smoothly" to the point in the middle. So the function is continuous at that point. Two things are important here: 1) that it is smooth[3] and 2) that it approaches to the same point from both sides.

And the generalization to $\mathbb{R}^n$ of this is given by the following definition, where the reader can observe the use of balls instead of one-dimensional intervals.

---

[3]Smooth curve has a proper definition in mathematics, but in this case please take only the intuitive idea of *smoothness.*

Definition 32. *Continuous Function on an Open Set. We say that $f : \mathbb{R}^n \to \mathbb{R}^n$ is continuous on $\hat{a} \in \mathbb{R}^n$ if and only if $\forall\ \epsilon > 0, \exists \delta > 0$ such that if $x \in B_\delta(\hat{a}) \implies f(x) \in B_\epsilon(f(\hat{a}))$.*

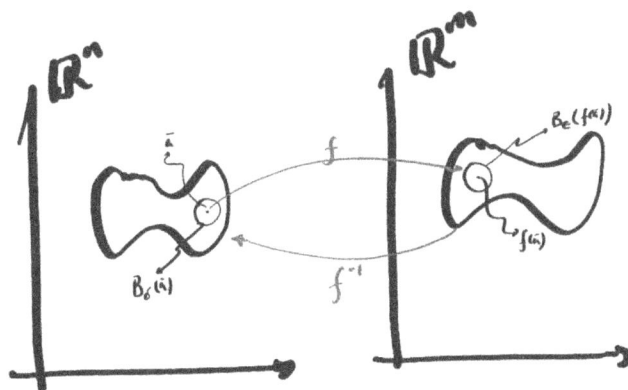

Now, the previous two definitions are equivalent, which also tells you a lot about how different areas of mathematics get together and form theory.

Also note that instead of intervals we use balls or circles; this is because of the many dimensions we are dealing with. In basic calculus, we only have the real numbers line; that is why we refer to intervals on the line; in two or more dimensions we need circles or spheres instead of intervals, in this way we cover any interval in any direction.

A very important theorem is the one that refers to the inverse image linked to continuity.

I will not prove it as the proof can be found in many classic textbooks and the proof will not add substantially to the aim of the book, but it is important for the reader to know the connection between topology and continuity.

Theorem 1. *Let $f : X \to Y$, also let $B \in Y$ be an open set. f is continuous if and only if the inverse image of B under f, $f^{-1}(B)$ is open.*

The proof can be found in classic analysis texts like [Rudin, 1974].

Definition 33. *Closed Set. Let X be a topological space. A subset F of X is closed if its complement F^c is open.*

Wait a minute! We defined a closed space based on an open space, but so far we have learnt that mathematics is based on definitions. If you notice the structure of the work so far it is based on formally defining concepts; in this way we can go ahead with rigorous mathematical logic. So a rightful questions from the reader would be: what is an open space? also, why is it that we need to define a closed space through an open space?

Open sets and open spaces are the finest tissues with which topological spaces are made of.

This is a very good place to introduce another important concept, which is *locality*. Many concepts and properties in mathematics start as local but then can be generalized.

When one studies topology one generally start with two concepts: one of open sets and the other of locality. The definitions of continuity, open sets and many of other topological concepts are local i.e. in a neighborhood. We then extend the local concepts and generalize for all space or other cases.

One of the golden nuggets of formation that mathematics teaches us is problem-solving techniques. In this case a "from particular to general and to particular again" technique works perfectly. We will see other problem solving lessons in the next chapters and how mathematical theory is built using these techniques.

It would be important to note that the *accordion theory*[4] discussed with David Semach and myself applies here to the full, reinforcing it in the case of building scientific knowledge.

But let's go back to our definition:

For a better understanding we will contain our definitions in the n-dimensional euclidean space $\mathbb{R}^n$, but the definitions (as said above) can be extended to more general cases or can be constrained to more particular ones.

One of the main definitions is the "metric". Please note that a metric is not the same as a measure. There is also a proper definition of measure that I will cover in this book. But first let us define the concept of *metric*.

---

[4]See chapter one for more on *accordion theory*.

Definition 34. *Metric. Let X be a set, we say that the function $d : X \times X \to \mathbb{R}$ is a metric on X the following properties hold $\forall x, y, z \in X$*

1. $d(x, x) \geq 0$

2. $d(x, y) = 0 \implies x = y$

3. $d(x, y) = d(y, x)$

4. $d(x, y) \leq d(x, z) + d(y, z)$ *(triangle inequality)*

We call function d a metric or a distance function and X a metric space.

Now, a metric is a distance and what this definition says is that the distance has to be always positive and that basically you need two points to measure it.

A special case of metric is the norm. The norm, which is generally associated to the length of a vector is nothing else but the distance between the initial and ending points. This is why, a norm which is applied to only one object is also a metric.

Most vector spaces are metric and normed spaces. This is because they have a distance function that follows the properties above for all the elements of the vector space. To support my definition of mathematics in the previous chapter, we can now give more examples of meaningful structures, like normed spaces and metric spaces.

For example $\mathbb{R}^2$ the two-dimensional Euclidean space is a vector space and a metric space under the euclidean 2-norm.

Please note that a "metric" is a foundational concept for topology and vector spaces and a "measure" is foundational for probability theory. It is also important to note that a metric is different to a measure. The first one is about distances between objects and the second one is about the sizes of the objects.

So let us define now what a norm is[5] .

Definition 35. *Norm. We say that V be is a "normed" vector space if there is a defined norm on V; and we say that $\|\| : V \to \mathbb{R}$ is a norm function of the Euclidean space if and only if the following properties hold:*

1. *$\|\bar{x}\| > 0$ and $\|\bar{0}\| = 0$*

   $\forall \bar{x} \in V, \bar{x} \neq \bar{0}$ where $\bar{0}$ is the additive identity.

2. *$\|\lambda \bar{x}\| = |\lambda| \|\bar{x}\|$*

   $\forall \bar{x} \in V$ and $\forall \lambda \in \mathbb{R}$

3. *$\|\bar{x} + \bar{y}\| \leq \|\bar{x}\| + \|\bar{y}\|$ (triangle inequality)*

---

[5]The definition of *measure* will be outlined and explained in the following chapters.

Another important operation in vector spaces is the inner product as it has a lot of important geometric and algebraic properties.

Definition 36. *Inner Product. Let V a vector space, we say that $\langle,\rangle : V \to \mathbb{R}$ is an inner product if and only if, $\forall \overline{x}, \overline{y}, \overline{z} \in V$ the following hold:*

     1. $\langle \overline{x}, \overline{y} \rangle = \langle \overline{y}, \overline{x} \rangle$

     2. $\langle \lambda \overline{x}, \overline{y} \rangle = \langle \overline{x}, \lambda \overline{y} \rangle = \lambda \langle \overline{x}, \overline{y} \rangle$

     3. $\langle \overline{x}, \overline{y} + \overline{z} \rangle = \langle \overline{x}, \overline{y} \rangle + \langle \overline{x}, \overline{z} \rangle$

     4. $\langle \overline{x} + \overline{y}, \overline{z} \rangle = \langle \overline{x}, \overline{z} \rangle + \langle \overline{y}, \overline{z} \rangle$

     5. $\langle \overline{x}, \overline{x} \rangle > 0 \forall \overline{x} \neq \overline{0}$

Using the norm $\|\|$ and the inner product $\langle,\rangle$ then we can finally define with formality the notion of "disk" of "ball":

Definition 37. *Ball. The ball with radius epsilon centered at $\overline{x}$ is defined as follows:*

$$B_\epsilon(\overline{x}) = \{\overline{y} \in \mathbb{R}^n \mid \|\overline{x} - \overline{y}\| < \epsilon\} \tag{4.1}$$

$\forall \epsilon \in \mathbb{R} > 0$

This definition gives us the idea of a set as an orange without the rind. This is crucial for the notion of continuity and in general for all infinitesimal calculus, as all the border of the ball should NOT be considered but only approached. Remember the notion of limit in basic calculus, in which we "approach" to the value but never get to the actual value. This is the same concept.

Now, sometimes we need to consider the border or "cover" of the ball, therefore, let us define it:

Definition 38. *Closed Ball. The closed ball with radius epsilon centered at $\overline{x}$ is defined as follows:*

$$\overline{B}_\epsilon(\overline{x}) = \{\overline{y} \in \mathbb{R}^n \mid \|\overline{x} - \overline{y}\| \leq \epsilon\} \tag{4.2}$$

$\forall \epsilon \in \mathbb{R} > 0$

Taking the analogy of the orange, this one would also consider the rind. And finally just to be thorough, there is another ball in topology: the hollow ball.

Definition 39. *Hollow Ball. The closed ball with radius epsilon centered at $\overline{x}$ is defined as follows:*

$$B'_\epsilon(\overline{x}) = \{\overline{y} \in \mathbb{R}^n \mid 0 < \|\overline{x} - \overline{y}\| < \epsilon\} \tag{4.3}$$

$\forall \epsilon \in \mathbb{R} > 0$

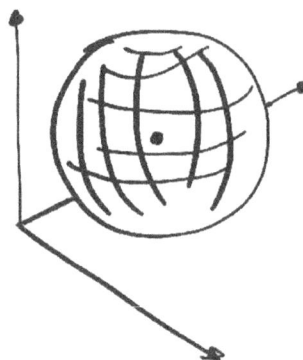

Note that the hollowed ball is an open ball too. And the analogy is that this ball does not have a center but it definitely has the "rind" or boundary.

These definitions will allow us to move towards more complex definitions like neighborhoods, interior points and boundary points and further concepts like closed and open sets and compact spaces.

All these concepts are crucial to read advanced mathematics texts and that is why I am outlining them.

Definition 40. *Interior Point. Let $X \in \mathbb{R}^n$ and let $\overline{x} \in X$; we say that $\overline{x}$ is an interior point of X if and only if $\exists \epsilon > 0$ such that $B_\epsilon(\overline{x}) \subset X$.*

This means that the ball has to be totally contained in the set and no part of the ball should be outside the set. Even in the extreme case when the ball is at the limit of the "rind" or boundary of the set, one could always build a ball totally contained in the set; why? Because the ball is open and we can always find a number as small as we want (we normally refer to it in mathematics as an *epsilon*); this number is little enough but positive such that the ball is always inside the set and no part is "sticking out".

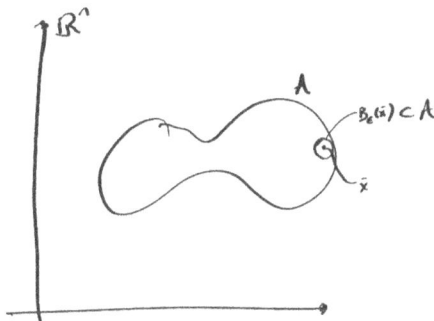

If for any reason the ball was not totally contained in the set or a bit of stuck out or touch the border of the set, then the point would not be an interior point. Given this, we can define:

Definition 41. *Interior of a Set. Let $X \in \mathbb{R}^n$, then $X^\circ = \{\overline{x} \in \mathbb{R}^n \mid \overline{x}$ is an interior point of $X\}$. X° is known as the interior of X.*

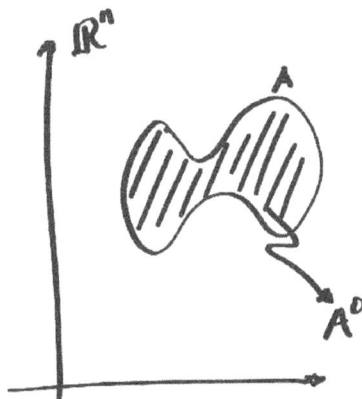

With this we can finally define an open set, which is one of the fundamental concepts not only in topology but in mathematics in general:

Definition 42. *Open Set. Let $X \subset \mathbb{R}^n$, it is said that X is an open set if and only if:*

$$\forall \overline{x} \in X, \exists \epsilon > 0 \ni B_\epsilon(\overline{x}) \subset X \tag{4.4}$$

This means that all the points of X are interior points; in other words, X is formed of interior points and only interior points.

Are all points of a set in the set? Well, no, not exactly. In mathematics we will make a distinction between the points on the boundary and all the other points in a set.

Definition 43. *Boundary Point. Let $X \subset \mathbb{R}^n, \overline{x} \in X$; we say that: $\overline{x}$ is a boundary point of X if and only if $\forall \epsilon > 0$*

$$B_\epsilon(\overline{x}) \cap X \neq \emptyset \tag{4.5}$$

and

$$B_\epsilon(\overline{x}) \cap X° \neq \emptyset \tag{4.6}$$

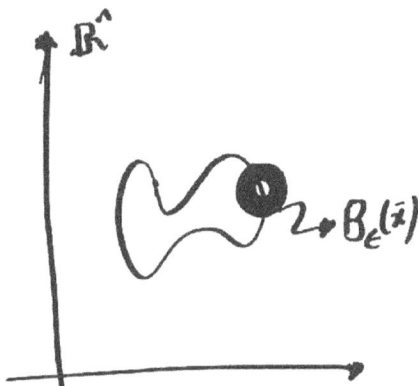

With this, we can now define the boundary of a set! Given that $X \in \mathbb{R}^n$ we have the following definition:

Definition 44. *Boundary.* $\partial X = \{\overline{x} \in \mathbb{R}^n \mid \overline{x} \text{ is an boundary point of } X\}$

∂X is known as the boundary of X; a point on a boundary is one such that one can build a ball -even the smallest possible one- such that this ball would touch the interior and exterior of the set at the same time.

Definition 45. *Exterior.* *Given that* $X \in \mathbb{R}^n$:

$$X^c = \{\overline{x} \in \mathbb{R}^n \mid \overline{x} \in (X \cup \partial X)^c\} \tag{4.7}$$

X^c is known as the complement of X and contains all the exterior points of X.

Therefore we have the following topological trichotomy; in other words, given a point $\overline{x} \in \mathbb{R}^n$ we have that only one of the following propositions hold:

1. $\overline{x} \in X$; this means that x is an interior point of X

2. $\overline{x} \in \partial X$; this means that x is a boundary point of X

3. $\overline{x} \in X^C$; this means that x is an exterior point of X

Do you see now how math works? This is one of the things mathematicians are really good at doing: Building a solid body of knowledge based on formal definitions and rigorous logic.

Two very important concepts that will be very useful to understand further ideas are the "accumulation point" and the "derived set".

Definition 46. *Accumulation Point. Let $X \subset \mathbb{R}^n$ and $\overline{x} \in X$. It is said that $\overline{x}$ is an accumulation point of X if and only if*

$$\forall \epsilon \in \mathbb{R}, B_\epsilon B'_\epsilon(\overline{x}) \cap X \neq \emptyset \tag{4.8}$$

This means that $\overline{x}$ is either an interior point or a boundary point of X. And given that $X \subset \mathbb{R}^n$, we define $X' = \{\overline{x} \in \mathbb{R}^n \mid \overline{x}$ is an accumulation point $\}$. This is known as the derived set of X (some authors call it *perfect set*). This means that X' or the derived of X is the interior of X and its boundary.

You could see that in order to define an interior point, we needed to define what a norm (or length or module) of a vector was and also we needed to define the inner product in a vector space.

Now you can see that the definition with which we started the section can be better understood now and even revisited:

Definition 47. *Closed Set Revisited. We say that $X \in \mathbb{R}^n$ is a closed set if and only if X^c is open; being $X^c = \mathbb{R}^n - A$. This definition is equivalent to the one stated before: Let X be a topological space. A subset F of X is closed if its complement F^c is open.*

A set X is closed if and only if X contains all of its accumulation points; in other words, $A' = A$.

This might be rather arid for some readers, but please bear with me, this is only a short intermission in our journey to understand homeomorphisms and ergodic transformations to build the foundations or the theory and have a common language and common understanding of the concepts and ideas.

So let us continue with a concept of the utmost importance. This concept is studied in linear algebra, analytic geometry, calculus and topology: vector space.

Definition 48. *Vector Space. A vector space is a set V with 2 binary operations, addition and multiplication by scalar where the following eight conditions hold for all elements in V and any scalars in $\mathbb{R}$.*

1. *Commutativity:* $\overline{x} + \overline{y} = \overline{y} + \overline{x}$

2. *Associativity under addition:* $(\overline{x} + \overline{y}) + \overline{z} = \overline{x} + (\overline{y} + \overline{z})$

3. *Additive identity:* $\forall \overline{z} \in V, \exists \overline{0} \in V$ *such that* $\overline{0} + \overline{x} = \overline{x} + \overline{0} = \overline{x}$

4. *Existence of additive inverse:* $\forall \overline{x}, \exists - \overline{x}$ *such that* $\overline{x} + (-\overline{x}) = \overline{0}$

5. *Associativity of scalar multiplication:* $r(s\overline{x}) = (rs)\overline{x}$

6. *Distributivity of scalar sums:* $(r + s)\overline{x} = r\overline{x} + s\overline{x}$

7. *Distributivity of vector sums:* $r(\overline{x} + \overline{y}) = r\overline{x} + r\overline{y}$

8. *Scalar multiplication identity:* $1\overline{x} = \overline{x}$

By now the reader will realize how mathematics works: a) definitions b) more definitions c) properties based on definitions d) theorems e) proofs f) corollaries g) proofs.

A very important part of a mathematician's work is to define. Concepts like action, or ball are not used loosely. In mathematics even these words have specific definitions. There is an old joke about a mathematician who had a complex relationship with the wife, which meant that he had a real and an imaginary part in the relationship (alluding to complex numbers).

As you can see, definitions are important so we understand each other in any language and in any discipline involving mathematics. It is really powerful to communicate in a common language and under the same understanding, because this is the way in which we can develop collaborative knowledge.

Definitions in mathematics are subject to scrutiny, and relevance, but certainly not subject to interpretation.

This is why, although this section might seem "dry", once it is understood will give a solid foundation to understand more structured concepts.

In topology we need to understand all this kind of points to see if they are in or out or on the boundary of spaces and if functions can cover or act on them. Another important type of point is the boundary point:

Definition 49. *Closure. If E is a subset of a topological space X, the closure of E is the intersection of all the closed subsets of X that contain E. The closure of E is denoted by $\bar{E}$.*

Definition 50. *Dense Set. Let X be a topological space and $D \in X$. We say that D is dense in X if $\bar{D} = X$. More generally it is said that D is dense in E if for any $E \in X, E \subset \bar{D}$.*

Definition 51. *Separable Set. A topological space in which a countable dense set exists is called separable.*

This is of the utmost importance when applying clustering algorithms as separability is a condition to have clear segmentation.

It is good to have a notion of separability but in this case we will properly define what a separable set is, so we can formalize the theory and understand other papers.

Definition 52. *Homeomorphism. Let X and Y be topological spaces and let $f : X \to Y$. We say that f is a homeomorphism of X on Y if F is bijective and F and F^{-1} are continuous.*

Definition 53. *Open Cover. Let X be a topological space and $E \subset X$. An arbitrary open cover of E is a collection $\mathcal{U} = \{U_\alpha : \alpha \in A\}$ of open subsets of X such that $E \subset \cup_{\alpha \in A} U_\alpha$. If $B \subset A$ the collection $\mathcal{V} = \{U_\alpha : \alpha \in B\}$ is called a subcover of U if V is a cover in itself.*

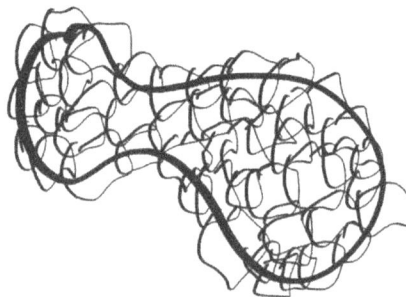

Definition 54. *Compact Set. Let X be a topological space. It is said that a subset K of X is compact if each open cover of K has a finite subcover.*

The classic theorem of Heine-Borel says that all compact sets in $\mathbb{R}^n$ are closed and bounded.

A vector space over the real numbers is an Abelian group with a scalar multiplication $R \times V \to V$ with the properties of a vector space as shown in definition 48.

If $Hom(V)$ is the group of homomorphisms of the additive group V and $\mathbb{R}^*$ is the multiplicative group of real numbers except zero, then the multiplication by scalars defines an action of $\mathbb{R}^*$ on $Hom(V)$ such that

$$(x + y)\alpha = x\alpha + y\alpha \tag{4.9}$$

This is the distributive property in terms of the additive structure in $\mathbb{R}$.

Definition 55. *Linear Transformation. Let V and W vector spaces. Then a mapping or transformation $T : V \to W$ is linear if $T(\alpha + \beta) = T(\alpha) + T(\beta)$ and $T(x\alpha) = xT(\alpha) \forall \alpha, \beta \in V$ and $x \in \mathbb{R}$.*

Definition 56. *Isomorphism. Let $T : V \to W$ a linear transformation. If T is bijective then we say that T is an isomorphism.*

Definition 57. *Isomorphic Spaces. Two vector spaces V and W are isomorphic if there is an isomorphism between them.*

It is known that if $V = W = \mathbb{R}^n$ then the linear transformations of $\mathbb{R}^n$ onto itself map with $M(n, \mathbb{R})$ and the isomorphisms with $\mathbb{GI}(n, \mathbb{R})$.

4.2   Notion of Algebraic Topology

Algebraic topology will help us identify invariants between topological spaces and homeomorphisms and will use algebraic resources to solve topology problems. One of the main concepts that will help us do this is *lifting*.

Although it was mentioned in the previous section it would be good to note at this point that when $T : V \to V$ it can happen that $T(\alpha) = x\alpha$ for some $x \in \mathbb{R}$. In this case we call α the eigenvector and x the eigenvalue.

Now, we will construct a homeomorphism on the circle. This will illustrate why topological spaces are necessary to express functions between vector spaces and how we intersect group theory with topology.

Let $\bigoplus : [0, 2\pi] \times [0, 2\pi] \to [0, 2\pi)$ such that $(\theta_1, \theta_2) \mapsto \theta_1 + \theta_2 (mod 2\pi)$ via $\bigoplus$ or $\theta_1 + \theta_2 = \theta_1 + \theta_2 (mod 2\pi)$ makes $[0, 2\pi)$ an additive isomorphic group via ϕ onto S^1. See commutative diagram after definition 14.

If $\phi[0, 2\pi) \to S^1$ is a bijection, then we define a new topology τ on $[0, 2\pi)$ that does not correspond to the one that the interval inherits as a subset of $\mathbb{R}$ defined by

$$\tau = \{\phi^{-1}(U) \mid U \in S^1\} \tag{4.10}$$

when U is open.

Then not only does ϕ becomes an isomorphism but a homeomorphism too.

A *lift or lifting* $f(x)$ of a transformation $T : S^1 \to S^1$ is a functions for which the following holds:

$$e^{2\pi i f(x)} = T(e^{2\pi i x}) \tag{4.11}$$

The lifting is useful because it keeps memory of the periodicity of the circle; in other words, with the lifting one can identify the number of times T "goes around" on the circle. In fact, the *Rotation Number Theorem* that we will analyze later will allow us to calculate the limit over f and not over T.

The formal and more general definition of a lifting is:

Definition 58. *Lifting. Let X be a topological space and let $\pi : E \to Y$ continuous with E and Y topological spaces. Let $f : X \to Y$ continuous. Then we say that $f : X \to E$ is a lifting of $g = T \circ \phi$ if the following diagram commutes.*

$$
\begin{array}{ccccc}
& & & & E \\
& & & \nearrow & \big\downarrow {\scriptstyle \phi = \pi} \\
& {\scriptstyle g=f} & & & \\
X & \xrightarrow{\ \ \phi \ \ } & S^1 & \xrightarrow{\ \ T \ \ } & Y
\end{array}
$$

In topology the notion of lifting is very general, and in a more open context f does not lift T but the composition of $T \circ \phi$.

But for the purpose of this book, we will not generalize in such way, but we will say that in particular f always lifts T and always exists.

To show that the lifting always exists one can refer to the fabulous book of "Algebraic Topology: A First Course" by Martin Greenberg [Greenberg, 1981].

It is worth noting that if T in continuous then f is continuous and that given T, there are as many liftings as integers $\mathbb{Z}$. In other words, if f is a lifting, then any $g = f + 2\pi n$ also is.

So far, we have seen how algebraic structures and topological spaces connect through algebraic topology (see all the commutative diagrams). Topology is important to understand continuity, but also to understand the spaces in which functions operate. Studying functions' context is important to know their possibilities, reaches and limits (remember functions in the examples of the XVII century scientists mentioned in chapter one).

In the next chapter we will connect algebra and topology with dynamical systems .

Chapter 5

Rotation Number: An Introduction to Iterations

> ... not only as to the things which I have explained, but also to those which I have intentionally omitted so as to leave to others the pleasure of discovery.

> *Discourse on Method*
> *René Descartes*

In this section I will start using a little bit heavier mathematical theory and you will see how I start linking concepts seen in the previous chapters, for example homomorphisms, liftings, congruence and basic concepts of calculus like limits. But most importantly, the reader will see how iterations work in mathematics applied to a specific problem.

Following the accordion and the Plata-Boere theories, I will firstly try to set the problem between the abstract and the concrete, this the reader will see how I will talk about speed, movement and distance which are concepts related to basic mechanics studied in middle school physics, but with an added complexity. And through iterations, I will define circles and most importantly, the concept of recurrence.

So, let us start: One of the most important parameters associated with the transformation on the circle is the rotation number. This number indicates the speed at which a rotation moves over S^1. Remember that a rotation can be seen as a dynamical system over a closed space.

As we will explain in the next section, if this number is irrational then the rotation does not have periodic points and it is uniquely ergodic. In fact, this number is calculated through a lifting associated to such transformations, but we say that the rotation number belongs to the transformation.

Before defining the rotation number, we need to define some concepts and emphasize some details of functions on the circle S^1 like 1-to-1 properties and image (or range) of functions.

The first observation is that for some interval of length less than 2π, and ϕ a function that preserves orientation and is also injective. In other words, if (a, b) is a small interval and $b - a < 2\pi$ is naturally directed from a to b; then the arc $\overset{\frown}{e^{ia}e^{ib}}$ is anti-clockwise.

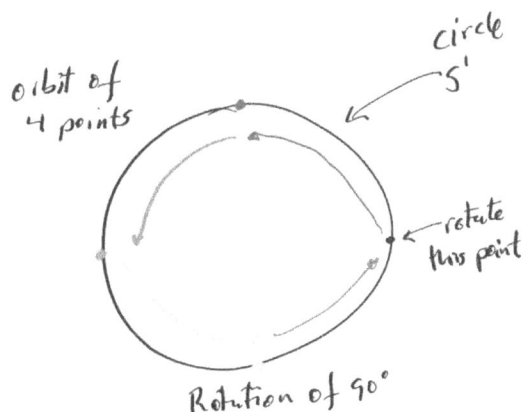

Another important observation is that the interval $[0, 2\pi)$ can be substituted by any interval of length 2π. Another classic option is $[-\pi, \pi]$. And if we iterate the directed arcs we will end up with defined orbits on the circle.

It is convenient to note that if f_1 and f_2 are liftings of T then the rotation number of f_1 is equal to the rotation number of $f_2 + m$ for a given $m \in \mathbb{Z}$. This is due to the actual definition of lifting, because for a given integer m, $e^{2\pi i m}$ recurs to the same point on the circle; this is the recurrence of S^1.

This assures that the rotation number does not depend on the lifting under translations.

In the following theorem a special type of lifting will be used, which is the one for which $f(x + 1) = f(x)$ holds. But as stated above the rotation number does not depend on the lifting.

Theorem 2 (Rotation Number). *For a homomorphism* $T : S^1 \to S^1$ *that preserves orientation and a lifting* f *that represents* T, *the limit*

$$\lim_{n \to \infty} \frac{f^n(x)}{n} = \alpha \qquad (5.1)$$

exists and does not depend on $x \in \mathbb{R}$. *The rotation number is rational if and only if the homomorphism* T *has a fixed point for a given iteration different to zero.*

The proof has a lot of good points that illustrate mathematical concepts, like the use of telescopic series, iterations, analysis, synthesis, problem decomposition and other calculus tools that are important to model, evaluate and solve problems in practice. Hence I will go over the proof as it is relevant to the spirit and aim of the book.

Proof. The proof is divided in three parts:

1. First we need to prove that if the limit exists for a specific point x_0, then it exists for any other point $x \in \mathbb{R}$.

2. Prove that the limit exists for a given point x_0 when T^k has fixed points.

3. Prove that the limit exists for T without fixed points with $x_0 = 0$ and by the result of point 1 of the proof the limit should exist for any point $x \in \mathbb{R}$.

5.1   Problem Solving by Limits

When dealing with decision science problems, a good technique is to find the limits of our problem. In practical terms, no problem is infinite (only theoretical ones are); so finding an upper bound and a lower bound is a good approach. If we set these parameters, then we limit our problem and thus finding solutions is easier.

Now, let us apply this concept to the proof: We know that a homomorphism T that preserves orientation in S^1 is of the form $T(x) = f(x) mod(1)$ in $0 \leq x \leq 1$, where $f(x)$ is a continuous increasingly monotone function on $\mathbb{R}$ and satisfies the condition that:

$$f(x+1) = f(x) + 1 \qquad (5.2)$$

Firstly supposed that the limit $\lim_{n \to \infty} \frac{f^n(x)}{n} = \alpha$ exists for some x_0 and take any arbitrary $x \in \mathbb{R}$ and a certain $m \in \mathbb{Z}$ such that:

$$x_0 + m \leq x < x_0 + m + 1 \qquad (5.3)$$

and given that f^n is increasingly monotone , we have that for any n:

$$f^n(x_0 + m) \leq f^n(x) < f^n(x_0 + m + 1) \qquad (5.4)$$

and

$$f^n(x_0) + m \leq f^n(x) < f^n(x_0) + m + 1 \qquad (5.5)$$

thus, we have:

$$m \leq f^n(x) - f^n(x_0) < m + 1 \qquad (5.6)$$

and dividing by n we get:

$$\frac{m}{n} \leq \frac{f^n(x)}{n} - \frac{f^n(x_0)}{n} < \frac{m+1}{n} \qquad (5.7)$$

In this expression we have that the sequence difference is bounded on the left and right. If we know the limit of the bounds and also we know the limit of one of the sequences, then we can say the following, taking the limit when $n \to \infty$:

$$\lim_{n \to \infty} \frac{m}{n} \leq \lim_{n \to \infty} \frac{f^n(x)}{n} - \lim_{n \to \infty} \frac{f^n(x_0)}{n} < \lim_{n \to \infty} \frac{m+1}{n} \qquad (5.8)$$

then

$$0 \leq \lim_{n \to \infty} \frac{f^n(x)}{n} - \alpha \leq 0 \qquad (5.9)$$

and

$$\alpha \leq \lim_{n \to \infty} \frac{f^n(x)}{n} \leq \alpha \tag{5.10}$$

Therefore, the limit does not depend on the initial condition due to the monotony of the lifting. And given that if the limit $\lim_{n \to \infty} \frac{f^n(x)}{n}$ exists for some $x_0 \in \mathbb{R}$, then it exists for any $x \in R$ and it is the same i.e. it does not depend on the point x taken.

This implies that

$$\lim_{n \to \infty} \frac{f^n(x) - f^n(x_0)}{n} = \alpha \tag{5.11}$$

In turn this means that if the limit $\lim_{n \to \infty} \frac{f^n(x)}{n}$ exists for some x_0 then it exists for any $x \in \mathbb{R}$ and does not depend on x_0.

With this we finish the first point of the proof.

5.2   Problem Solving by Elimination

The objective of the second part of the proof is to show that the limit exists for a given x_0, being one of the two cases that T^k has fixed points.

To do this we will apply telescopic series which normally simplify when we cancel terms out by operating them. In the end we end up with a concise term that reflects the entire process or problem. Synthesizing problems in this way is a good approach to problem solving and moreover it helps the abstraction of complex processes to their major components. Identifying the most important component in a process is also very useful in design thinking. Alec Boere has been successfully using this technique when leading design thinking workshops.

Now, let's suppose that $T^k(x_0) = x_0$; this means that the kth iteration of T falls again on x_0; therefore if T^k has a fixed point, then T has a periodic point and vice versa, if T has a periodic point, the T^k has a fixed point.

A lifting of T^k that is very convenient for the proof is $f^k(x_0) = x_0 + r$, with $r \in \mathbb{Z}$. r is called the "winding" number and it is precisely the constant that measures the recurrence of T on the circle; in other words, it is counting the number of rounds that the dynamical system T^k does on the circle S^1.

This lifting was chosen in a way that r measured the number of rounds that T^k goes on the circle; in fact one can chose a lifting that does not measure it, for example if a lifting is $f(x) = x + r$, then another one can be $f(x) = x + 800$; the latter does not measure the number of rounds that T^k goes on S^1.

Then the lifting of T^k is $f^k(x_0) = x_0 + r$ with $r \in \mathbb{Z}$ and for any $l \in \mathbb{Z}$ we find that any multiple of k or lk, and any multiples of the kth iteration are as follows:

For $l = 0$:

$$f^{0k}(x_0) = f^0(x_0) = x_0 \tag{5.12}$$

For $l = 1$:

$$f^{1k}(x_0) = f^k(x_0) = x_0 + r \tag{5.13}$$

For $l = 2$:

$$f^{2k}(x_0) = f^{k+k}(x_0) = f^k(f^k(x_0)) = f^k(x_0 + r) \tag{5.14}$$

$$= x_0 + r + r = x_0 + 2r \tag{5.15}$$

For $l = 3$:

$$f^{3k}(x_0) = f^{k+k+k}(x_0) = f^k(f^k(f^k(x_0))) = \tag{5.16}$$

$$= f^k(f^k((x_0 + r)) = f^k(x_0 + 2r) = x_0 + 2r + r = x_0 + 3r \tag{5.17}$$

and in general

$$f^{1k}(x_0) = x_0 + lr \qquad (5.18)$$

Proof (by induction over l)

1. it is valid for $i = 1$

   So if $l = 1$ we obtain

$$f^{1k}(x_0) = f^k(x_0) = x_0 + r \qquad (5.19)$$

2. Assume valid for n i.e. $f^{nk}(x_0) = x_0 + nr$ and

3. Show that $f^{n+1)k}(x_0) = x_0 + (n+1)r$:

   If $f^{nk} = x_0 + nr$, then $f^k(f^{nk}(x_0)) = f^k(x_0 + nr + r = x_0 + (n+1)r$

   Hence, if the expression $f^{lk}(x_0) = x_0 + lr$ is divided by lk we have:

$$\frac{f^{1k}(x_0)}{lk} = \frac{x_0 + r}{lk} = \frac{x_0}{lk} + \frac{lr}{lk} \qquad (5.20)$$

   and if we take the limit when $l \to \infty$, then

$$\lim_{l \to \infty} \frac{f^{1k}(x_0)}{lk} = \lim_{l \to \infty} \frac{x_0}{lk} + \lim_{l \to \infty} \frac{lr}{lk} = \frac{r}{k} \qquad (5.21)$$

therefore

$$\lim_{l \to \infty} \frac{f^{1k}(x_0)}{lk} = \frac{r}{k} \tag{5.22}$$

Up to this point an important result is that for T^k with a fixed point on some iteration $k \neq 0$ and for its multiples, the limit exists and it is rational. The interpretation of this result is:

If $f^k(x_0) = x_0 + r$ then $f^k(x_0) - x_0 = r$, in other words, r is the distance between the kth iteration of f starting on x_0 and then we can say that the rate at which each iteration of f advances at each iteration is $\frac{r}{k}$.

If the function covered a distance r in k iterations, then $\frac{r}{k}$ is the average distance that f took per iteration until T passes through the same point (remember that k is the period).

To conclude, $\frac{r}{k}$ is the speed at which T goes around and returns to a periodic point, and although T does not keep record ("or has any memory") of the number of rounds, f does!

It is important to note that $\frac{r}{k}$ is the ratio between k and the point x_0 and not to the origin.

Up to now, I have analyzed the case of the multiples of k, so, next we have the case of any integer n.

To do this, I will use the division algorithm, that allows to express any integer n as $n = lk + s$ with $0 \leq s < k$, i.e. n can be expressed as a multiple of k plus a number between 0 and k in order to avoid becoming the next multiple. Therefore:

$$f^n(x_0) = f^{lk+s}(x_0) = f^s(f^{lk}(x_0)) = f^s(x_0 + lr) \tag{5.23}$$

$$= f^s(x_0) + lr \tag{5.24}$$

Using the previous equations, if

$$f^n(x_0) = f^s(f^{lk}(x_0)) = f^s(x_0) + lr \tag{5.25}$$

then,

$$f^s(f^{lk}(x_0)) - f^s(x_0) = lr \tag{5.26}$$

and

$$f^{lk}(f^k(x_0)) - f^s(x_0) = lr \tag{5.27}$$

This last expression reduces the case of any integer to the case of the multiples of k and establishes -reassuring the result of the first part of the proof- that it does not matter the point where it is evaluated, the limit exists and does not depend on the given point.

One thing we can observe from this last result is that the distance between the kth iteration and its multiples are the same: lr.

Thus, if $f^k(x_0) = y$ with $y \in \mathbb{R}$, then the expression $f^{lk}(f^s(x_0)) - f^s(x_0) = lr$ can be written as:

$$f^{lk}(y) - y = lr \tag{5.28}$$

We must clarify that y depends on s and that it is not another point x_0. In other words, y is a product of iterating f s times.

Then dividing by lk we obtain:

$$\frac{f^{lk}(y)}{lk} - \frac{y}{lk} = \frac{lr}{lk} \tag{5.29}$$

and taking the limit when $l \to \infty$ we have:

$$\lim_{l \to \infty} \frac{f^{lk}(x_0)}{lk} - \lim_{l \to \infty} \frac{y}{lk} = \lim_{l \to \infty} \frac{lr}{lk} = \frac{r}{k} \tag{5.30}$$

On the other hand, if $l \to \infty$ then $n \to \infty$ too. And it the previous result is divided by n, then we get the following expression:

$$\frac{f^{lk}(y)}{n} - \frac{y}{n} = \frac{lr}{lk} \tag{5.31}$$

and

$$\lim_{n \to \infty} \frac{f^{1k}(y)}{n} - \lim_{n \to \infty} \frac{y}{n} = \lim_{n \to \infty} \frac{lr}{n} \tag{5.32}$$

But due to the fact that $n = lk + s$ and when $n \to \infty$ then $l = \frac{n-s}{k} \to \infty$; in addition, s and k are fixed, hence:

$$\lim_{n \to \infty} \frac{f^{1k}(y)}{n} - \lim_{n \to \infty} \frac{y}{n} = \lim_{n \to \infty} \frac{lr}{n} \tag{5.33}$$

can be expressed as:

$$\lim_{n \to \infty} \frac{f^{1k}(y)}{n} - 0 = \lim_{l \to \infty} \frac{lr}{lk + s} \tag{5.34}$$

and as y is fixed, then:

$$\lim_{l \to \infty} \frac{lr}{lk + s} = \lim_{l \to \infty} \frac{r}{k + \frac{s}{l}} = \frac{r}{k} \tag{5.35}$$

It is important to note that given that s remains constant, then the limit tends to zero as l approaches to infinity. And therefore:

$$\lim_{n \to \infty} \frac{f^n(x_0)}{n} = \frac{r}{k} \tag{5.36}$$

In summary:

$$\frac{f^n(x_0)}{n} = \frac{f^s(x_0)}{n} + \frac{lr}{lk + s} \to \frac{r}{k} \tag{5.37}$$

when $n \to \infty$

In order to conclude this second part of the proof, one has to highlight that the limit in question is the same as in the case of the multiples of k; in other words, the limit neither depends on the multiplicity of k nor on the point $x \in \mathbb{R}$. This confirms again the result of the first part of the proof.

So far we have interpreted the results in terms of the distance that f runs from the initial point. The next questions is what happens with the distances between f^k and f^{2k} and subsequently what happens with the distances between f^{2k} and f^{3k} etc.

We know that

$$f^k(x_0) - x_0 = r \tag{5.38}$$

then

$$f^{2k}(x_0) - f^k(x_0) = x_0 + 2r - x_0 - r = r \tag{5.39}$$

and

$$f^{3k}(x_0) - f^{2k}(x_0) = x_0 + 3r - x_0 - 2r = r \tag{5.40}$$

Hence, r is also the distance between each multiple of the iterations of f and therefore the geometric interpretation of the limit $\frac{r}{k}$ is the ratio at which f runs on each k iterations; the initial point has no consequence on this.

With this remark, the second part of the proof is finished. Now the last part of the proof treats the case in which T does not have fixed points.

5.3   Problem Solving by Iterations

For this third part, we will proof that the limit exists for T with no periodic points. In order to do this, we will assume that T does not have fixed points in any of its iterations; then $f^k(x) - x$ is not an integer. Remember that in the previous case $f^k(x) - x = r$ and r was an integer, and thus for all $x \in \mathbb{R}$ we have that:

$$x + r < f^k(x) < x + r + 1, r \in \mathbb{Z} \tag{5.41}$$

This means that the distance between $f^k(x)$ and x is not an integer (like in the previous case). $f^k(x) - x$ is between r and $r+1$ with r an integer. This means that the distance between $f^k(x)$ and x is bounded, then $r < f^k(x) - x < r+1$.

Now, let us choose a natural number k and apply the inequality $x + r < f^k(x) < x+r+1$ for the points $x = 0$, $f^k(0)$, $f^{2k}(0)$, $f^{3k}(0)$, ..., $f^{(n-1)k}(0)$ getting:

$$x + r < f^k(x) < x + r + 1 \qquad (5.42)$$

then

$$r < f^k(x) - x < r + 1 \qquad (5.43)$$

and for a specific point $x_0 = 0$ we have that:

$$r < f^k(0) - 0 < r + 1 \qquad (5.44)$$

which implies that:

$$r < f^k(0) < r + 1 \qquad (5.45)$$

In other words, the distance between $f^k(0)$ and the origin is between r and $r + 1$. Then the next question is what happens with the distance between $f^{2k}(0)$ and $f^k(0)$, or $f^{2k}(0) - f^k(0)$.

We know that

$$f^{2k}(0) = f^k(f^k(0)) \tag{5.46}$$

then

$$f^{2k}(0) - f^k(0) = f^k(f^k(0)) - f^k(0) \tag{5.47}$$

Now for the first iteration let us propose a substitution, let $y_1 = f^k(0)$, then $f^{2k}(0) = f^k(f^k(0)) = f^k(y_1)$ and

$$f^{2k}(0) - f^k(0) = f^k(f^k(0)) - y_1 \tag{5.48}$$

for which the following inequality holds:

$$r < f^k(y_1) - y_1 < r + 1, \forall y_1 \in \mathbb{R} \tag{5.49}$$

therefore

$$r < f^{2k}(0) - f^k(0) < r + 1 \tag{5.50}$$

For the case in which we want to calculate the distance between $f^{3k}(0)$ and $f^{3k}(0$, we only have to apply the same results and the same argument than in the previous case, thus we have:

$$f^{3k}(0) - f^{2k}(0) = f^k(f^{2k}(0)) - f^{2k}(0) \tag{5.51}$$

Now for the second iteration let us propose another substitution, let $y_2 = f^{2k}(0)$, then

$$f^{3k}(0) - f^{2k}(0) = f^k(y_2) - y_2 \tag{5.52}$$

and

$$r < f^{3k}(0) - f^{2k}(0) < r + 1 \tag{5.53}$$

and in general

$$r < f^{nk}(0) - f^{(n-1)k}(0) < r + 1 \tag{5.54}$$

or

$$f^{(l)k}(0) + r < f^{(l+1)k}(0) < f^{(l)k}(0) + r + 1, l = 0, 1, 2, ..., n - 1 \quad (5.55)$$

This means that the distance between any two consecutive multiples of the iterations of f will always be between r and $r + 1$. Then the sum of all the inequalities (n inequalities)

$$r < f^k(0) < r + 1$$

$$r < f^{2k}(0) - f^k(0) < r + 1$$

$$r < f^{3k}(0) - f^{2k}(0) < r + 1$$

$$\vdots$$

$$r < f^{nk}(0) - f^{(n-1)k}(0) < r + 1 \quad (5.56)$$

is

$$nr < f^{nk}(0) < n(r + 1) \quad (5.57)$$

Now, dividing by nk we obtain:

$$\frac{nr}{nk} < \frac{f^{nk}(0)}{nk} < \frac{n(r+1)}{nk}$$

$$\frac{r}{k} < \frac{f^{nk}(0)}{nk} < \frac{(r+1)}{k} \tag{5.58}$$

and applying the same argument for the first inequality we have:

$$\frac{r}{k} < \frac{f^{k}(0)}{k} < \frac{(r+1)}{k} \tag{5.59}$$

In other words:

$$\frac{r}{k} < \frac{f^{nk}(0)}{nk} < \frac{r}{k} + \frac{1}{k} \tag{5.60}$$

and

$$\frac{r}{k} < \frac{f^{k}(0)}{k} < \frac{r}{k} + \frac{1}{k} \tag{5.61}$$

Now, if we multiply this last expression by -1, we obtain

$$-\frac{r}{k} - \frac{1}{k} < -\frac{f^k(0)}{k} < -\frac{r}{k} \tag{5.62}$$

and adding them we get:

$$-\frac{1}{k} < -\frac{f^{nk}(0)}{nk} - \frac{f^k(0)}{k} < \frac{1}{k} \tag{5.63}$$

therefore

$$|\frac{f^{nk}(0)}{nk} - \frac{f^k(0)}{k}| < \frac{2}{k} \tag{5.64}$$

One of the original contributions to this work is the geometric interpretation of the rotation number under ergodic transformations. My background is on differential topology, so geometric abstraction is one of the things I enjoy most, hence the following note:

Now, if we analyze these inequalities geometrically, we have that the nk^{th} iteration of zero is between $\frac{r}{k}$ and $\frac{r}{k} + \frac{1}{k}$ and in the same way, the kth iteration of f on zero is therefore, the distance between each other is less than $\frac{1}{k}$ or:

$$|\frac{f^{nk}(0)}{nk} - \frac{f^k(0)}{k}| < \frac{1}{k} \tag{5.65}$$

Which is a stronger constraint than manipulating the inequality analytically. And this is an example of the power of geometric analysis. This is important although for the purpose of the proof it is enough to say that $| \frac{f^{nk}(0)}{nk} - \frac{f^k(0)}{k} |$ is less than $\frac{2}{k}$.

Thus, given that n and k are arbitrary values, they are interchangeable and while adding the two inequalities we can repeat the argument for k instead of n getting:

$$| \frac{f^{nk}(0)}{nk} - \frac{f^k(0)}{n} |< \frac{1}{n} \tag{5.66}$$

If the distance between $\frac{f^{nk}(0)}{nk}$ and $\frac{f^k(0)}{n}$ is less than $\frac{1}{n}$ the same applies for the distance between $\frac{f^{nk}(0)}{nk}$ and $\frac{f^k(0)}{k}$, which is less than $\frac{1}{k}$, then at most:

$$| \frac{f^n(0)}{n} - \frac{f^k(0)}{k} |< \frac{1}{n} + \frac{1}{k} \tag{5.67}$$

What we want to see is that the limit $\lim_{n \to \infty} \frac{f^n(x)}{n} = \alpha$ exists. In this case:

$$\lim_{n,k \to \infty} | \frac{f^n(0)}{n} - \frac{f^k(0)}{k} |< \frac{1}{n} + \frac{1}{k} = 0 \tag{5.68}$$

If we see $\frac{f^n}{n}$ as a sequence, we only need to prove that it is a Cauchy sequence to see that it converges and then the limit exists; let us remember that a series x_n is a Cauchy sequence if $\forall \epsilon > 0$ there is an $N \in \mathbb{N}$ such that for all n and k with $n > N$ and $k > N$, then the distance $\rho(x_n, x_k) < \epsilon$; i.e. $\mid x_n - x_k \mid < \epsilon$ or in our case, if we take the limit it should be clear that this condition holds:

$$\lim_{n,k \to \infty} \mid \frac{f^n(0)}{n} - \frac{f^k(0)}{k} \mid < \lim_{n,k \to \infty} \frac{1}{n} + \lim_{n,k \to \infty} \frac{1}{k} = 0 \qquad (5.69)$$

And because of the first result of the proof, if the limit exists for the case in which $x = 0$ then it exists for any $x \in \mathbb{R}$, therefore,

$$\lim_{n \to \infty} \frac{f^n(x)}{n} = \alpha \qquad (5.70)$$

exists.

Now, it is only left to prove that if α is rational, then at least one iteration of T different from zero has a fixed point.

Firstly assume that $\alpha = 0$. Then we need to prove that T has a fixed point. We will do this by *Reductio ad Absurdum*[1]: Suppose that the point does not exist; then $f(x) - x \neq 0$ for some x and therefore we can assume that $f(x) > x$ for some $x \in \mathbb{R}$. In particular $f(0) > 0$ and therefore $f^n(0) > f^{n-1}(0) > ... > 0$ because f is monotonic.

---

[1]Also known as *by contradiction*.

Then $f^n(0)$ is a monotonically increasing sequence and even more, $f^n(0) < 1$ for all n. Indeed, had we for some n_0 that $f^{n_0}(0) \geq 1$ then $f^{2n_0}(0) \geq f^{n_0}(1) = f^{n_0}(0) + 1 \geq 2$ and in general we would have that $f^{kn_0}(0) \geq k$ then $\frac{f^{kn_0}(0)}{kn_0} \geq \frac{1}{n_0}$ which contradicts the assumption that $\alpha = 0$.

Then the sequence $f^n(0)$ is monotonic and bounded. Now suppose that $x_0 = \lim_{n \to \infty} f^n(0)$ then

$$f(x_0) = \lim_{n \to \infty} f(f^n(0)) = \lim_{n \to \infty} f^{n+1}(0) = x_0 \qquad (5.71)$$

This means that x_0 defines a point on the circle which is fixed with respect to T.

Now suppose that $\alpha \in \mathbb{Q}$ with $\alpha = \frac{r}{k}$, then the function $g(x) = f^k(x) - r$ is a lifting of T^k, moreover:

$$\lim_{n \to \infty} \frac{g^n(x)}{n} = \lim_{n \to \infty} \frac{f^{kn}(x)}{n} - r = \lim_{n \to \infty} \frac{f^{kn}(x)}{kn} - r = 0 \qquad (5.72)$$

Therefore there is a fixed point with respect to T^k.

∎

As additional observations to the theorem we have that: If T does not have periodic points, then

$$| \frac{f^n(0)}{n} - \frac{f^k(0)}{k} | < \frac{1}{n} + \frac{1}{k} = 0 \qquad (5.73)$$

The distance between one iteration and the other is never the same, even more, each iteration is closer to each other every time and the limit:

$$\lim_{n,k \to \infty} | \frac{f^n(0)}{n} - \frac{f^k(0)}{k} | < \lim_{n,k \to \infty} \frac{1}{n} + \lim_{n,k \to \infty} \frac{1}{k} = 0 \qquad (5.74)$$

This means that for any point P_0 on the orbit of T there is another point P_1 on the orbit of T arbitrarily close to P_0. Therefore the orbit of T is dense on S^1.

If the limit is an irrational number, then the rotation T has a dense orbit.

Definition 59. *Rotation Number. If $T : S^1 \to S^1$ is an orientation-preserving homeomorphism and f is a function that represents it, we say that the rotation number:*

$$\alpha = \alpha(T) = \lim_{n \to \infty} \frac{f^n(x)}{n} (mod(1)) \qquad (5.75)$$

is the rotation number of T for $x \in \mathbb{R}$

For the case of superior dimensions, starting with the case of $\mathbb{R}^2$, we have that the two-dimensional torus (Tor^2), defined as $S^1 \times S^1$, holds the properties of each circle. Thus, each coordinate of the torus is represented by points of the form $(\theta_1 + 2\pi r, \theta_2 + 2\pi s)$ with r and $s \in \mathbb{Z}$. The liftings in this case are as follows:

$$
\begin{array}{ccc}
\mathbb{R}^2 & \xrightarrow{\; f=(f_1,f_2) \;} & \mathbb{R}^2 \\
{\scriptstyle (e^{2\pi ix},e^{2\pi iy})} \downarrow & & \downarrow {\scriptstyle (e^{2\pi ix},e^{2\pi iy})} \\
Tor^2 & \xrightarrow{\; T \;} & Tor^2
\end{array}
$$

The two-dimensional torus is very special because one can actually see it and because it can be taken as the set of all the equivalence classes of the points of the plan $\mathbb{R}^2$. To do this we need to analyze the torus from the plane as follows:

We create a grid of unit squares on the plane, starting by the square $C_1 = [0,1] \times [0,1]$; in other words, the square of all points in $\mathbb{R}^2$ that hold with $0 \leq x < 1$ and $0 \leq y < 1$.

And thus for any point $(x, y) \in C_1$ is considered the same as other point $(x', y') \notin C_1$ if:

$$x = x' + R(mod(1)) \tag{5.76}$$

$$y = y' + S(mod(1)) \tag{5.77}$$

with $R, S \in \mathbb{Z}$.

If R and S were less than 1, then the point (x', y') would be in C_1; i.e. $(x', y') \in C_1$.

In this case, the line $y = 1$ is associated with the line $y = 0$ and it is considered the same; this is why, when define the unit squares we take the strict equality at zero. similarly, the boundary points of the square that are taken into consideration for the case of x are $x = 1$ and not $x = 0$.

Then we "fold" the square to make it into a cylinder, and then we put together the top and bottom to make it a torus.

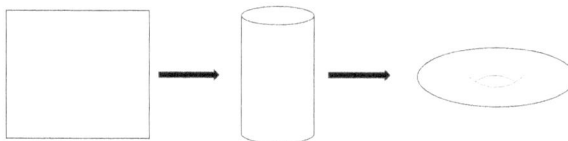

This torus can give a lot of information about the dynamic of systems in the plane without the need to analyze them in the plane itself, as this would be far more complicated.[2]

In the same way as in the previous case of the circle S^1, the calculations about some T on the torus will be done $mod(1)$.

---

[2]A great application of this concept can be found in [Plata, 2020] where a supermarket store is mapped to a torus.

For example, take the transformation $T : S^1 \to S^1$ such that one of its liftings f is $f(\bar{x}) = f(x, y) = (2x + y, x + y)$, that in the form of a matrix can be written as:

$$\begin{bmatrix} 2 & 1 \\ 1 & 1 \end{bmatrix}$$

That applied to a vector $\bar{x}$ is:

$$\begin{bmatrix} 2 & 1 \\ 1 & 1 \end{bmatrix} \begin{bmatrix} x \\ y \end{bmatrix} = \begin{bmatrix} 2x & y \\ x & y \end{bmatrix}$$

Now suppose that we take the point $x_0 = (\frac{1}{2}, \frac{1}{2})$ as the initial point, then $f(x_0) = f((\frac{1}{2}, \frac{1}{2}))$:

$$\begin{bmatrix} 1 & \frac{1}{2} \\ \frac{1}{2} & \frac{1}{2} \end{bmatrix} = \begin{bmatrix} \frac{1}{2} \\ 0 \end{bmatrix}$$

And $f^2(x_0) = f((\frac{1}{2}, 0))$:

$$\begin{bmatrix} 1 & 0 \\ \frac{1}{2} & 0 \end{bmatrix} = \begin{bmatrix} 0 \\ \frac{1}{2} \end{bmatrix}$$

$f^3(x_0) = f((0, \frac{1}{2}))$:

$$\begin{bmatrix} 0 & \frac{1}{2} \\ 0 & \frac{1}{2} \end{bmatrix} = \begin{bmatrix} \frac{1}{2} \\ \frac{1}{2} \end{bmatrix}$$

$f^4(x_0) = f((\frac{1}{2}, 0))$:

$$\begin{bmatrix} 1 & \frac{1}{2} \\ \frac{1}{2} & \frac{1}{2} \end{bmatrix} = \begin{bmatrix} \frac{1}{2} \\ 0 \end{bmatrix}$$

Therefore, this point has period 3 under T. One has to notice only that $T^4 = T$.

In the particular case that each one of the transformations acts independently on S^1, we get that the transformations of the torus are a system of the form $T_1 \times T_2$.

For example, similarly to the things pointed out in this section with respect to the circle S^1, $f(\bar{x}) = f(x) = (x + 1r, y + 1s)$, which is the identity on the torus.

If T has any periodic point $\bar{x}_0 = (x_0, y_0)$ then we have that $T^k(\bar{x}_0) = T^k(x_0, y_0) = (x_0, y_0) + (r, s)$.

Then each component of the transformation $T : T_1 \times T_2$ has its rotation number $\alpha(T_1)$ and $\alpha(T_2)$, then the ratio of the rotation numbers $\frac{\alpha(T_1)}{\alpha(T_2)}$ is the rotation number of the two-dimensional torus Tor^2. If the ratio is rational, the lines corresponding to the transformation are recurrent over themselves on the torus.

Chapter 6

Recurrence and Ergodicity:
An Introduction to Analysis

> I need only foresee the exceptions to the norm
> and calculate the most probable
> combinations.

Invisible Cities
Italo Calvino

This section will cover general topics of general advanced analysis. In mathematics advanced real analysis comprises basically measure theory (which is the basis of probability) and integrable functions. However I will introduce concepts of functional analysis like characters and endomorphisms and link them to measurable functions and spaces through ergodic transformations.

By the end of this chapter I will link the foundations of measure theory with the foundations of probability. In fact probability and statistics exist only within the measure theory space.

A very important concept in measure theory is that of a *signma-algebra* which is closely linked to Borel sets and Lebesgue measures.

In simple terms, the connection comes as follows: a Borel set is any set in a topological space (this was defined in chapter 3) built out of the countable unions and intersections of open sets (remember the concept of cover in chapter 3). And a set of Borel sets is a sigma-algebra.

Finally, and following the theory proposed by Boere and myself, I will construct a probability measure that will be defined by iterating a function over a space.

This is a clear example of the *Plata-Boere third iteration recurrent theory (THIRTY)*, in which we start by iterating functions in the classic dynamical systems theory and method, but end up by defining a probability.

6.1   Measurable Spaces

Ergodic theory studies the dynamic of measurable spaces. The sets and functions of interest are precisely the measurable ones. To study this we require a general framework about measure theory. Hence the following formal definitions that constitute the foundations of probability.

Definition 60. *Sigma Algebra. Let X be a set, we say that a sigma algebra σ-algebra of subsets of X is a collection $\mathcal{B}$ of subsets of X that satisfy the following conditions:*

1. *$X \in \mathcal{B}$*

2. *if $B \in \mathcal{B}$ then $B^c \in \mathcal{B}$*

3. *if $B_n \in \mathcal{B}$ for $n \geq 1$, then $\cup_{n=1|}^{\infty} B_n \in \mathcal{B}$*

Definition 61. *Measure Space. The pair $(X, \mathcal{B})$ is called a measurable space, and the members of $\mathcal{B}$ are called measurable sets on X*

Definition 62. *Measurable function. If X is a measurable space, Y is a topological space and $f : X \to Y$, then it is said that f is measurable if $f^{-1}(V)$ is a measurable set on X for all open sets $V \in Y$.*

Definition 63. *Finite Measure. A finite measure in the measurable space $(X, \mathcal{B})$ is a function $m : \mathcal{B} \to \mathbb{R}^+$, that satisfy:*

1. *$m(\emptyset) = 0$*

2. *$m(\cup_{n=1|}^{\infty} B_n) = \sum_{n=1}^{\infty} m(B_n)$ as long as B_n is a sequence of members of $\mathcal{B}$ disjoint one-to-one*

Definition 64. *Exterior Measure. For each set E of real numbers consider a collection of open intervals $\{I_n\}$ that cover E; in other words, a collection for which $E \subset \cup I_n$ and consider for each collection the sum of the lengths of their intervals. It is important to note that the lengths are positive numbers. Then we define the exterior measure $\mu^*(E)$ of E as the infimum of those sums. Using notation:*

$$\mu^*(E) = \inf_{E \subset \cup I_n} \sum l(I_n) \tag{6.1}$$

From the definition of μ^* it follows immediately that $\mu^*(\emptyset) = 0$ and if $E \subset F$ then $\mu^*(E) < \mu^*(F)$. It also follows that for a set consisting of only one point, the measure is zero.

An important proposition about the exterior measure is the following:

Proposition. The exterior measure of an interval is its length.

The reader can find a longer explanation of this in several places in the literature; I prefer the classic text [Royden, 1968, p.54]. In general the notion of measure is the length or width of something; what I am doing here is formalizing the concept so we can launch more ideas from it.

In summary this result allows us to connect the theory with its applications in a logical way. The reader will notice that the measure is actually the length, which makes sense in the "real world"[1].

---

[1]Remember that an interval is uni-dimensional.Therefore the length is the size of the interval

Definition 65. *Borel Algebra. Let X be a topological space. The σ-algebra generated by the closed subsets of X is called Borel σ-algebra, which is the calls of Borel subsets of X. Another way to phrase this is that given X then there is a minimum σ-algebra $\mathcal{B}$ in X such that each open set X belongs to $\mathcal{B}$. The members of $\mathcal{B}$ are called Borel sets on X.*

A σ-algebra is the essence and foundation of probability. The formal study of probability starts with this concept. What we learn about probability being the ratio between two numbers i.e. the probability of some event A occurring, denoted by $P(A) = \frac{\#PositiveOutcomes}{\#TotalOutcomes}$ is a special case which might not necessarily contemplate all possible cases (because it is intuitive).

For example when we ask what is the probability of getting tails when tossing a non-biased coin. The notional answer is 0.5. This is because you are counting two possible outcomes: head or tails, both with the same probability. But have you taken into account the possibility of the coin landing on its edge? That is certainly a possibility, right? (maybe low, but possible). So what would be the probability now?

Another example is when we are trying to join events. In the case of answering what is the probability of being female and studying mathematics, we need to have in our sigma algebra those two events (being female and studying mathematics). And the intersection should be one of the countable sets in it. The same would happen with any other operation: union, complement, etc.

In this chapter I will link measures with probability and moreover link the basic notions with the formal theory, so it will make sense.

This is constructing probability via iterations.

Now I will enunciate a theorem that states the existence of a measure that will be very useful when working on many cases involving probabilities, and even when working on dynamical systems, precisely because it is an invariant under rotations on the circle. And even when it is not the complete version, it is enough to understand the dynamic of discrete systems which are highly modeled in computer science.

What I recommend in this specific case (as an exception) is to use the results of the theorem more than the methods used in the proof. With this I am not dismissing the importance of the arguments, but for the purpose of this book and its continuity it is more relevant to concentrate on the concepts expressed in the theorems more than in the logical sequence of the proofs[2].

Theorem 3. *Let G be a topological compact group. Then there is a unique measure μ defined in the Borel σ-algebra $\mathcal{B}$ of subsets of G such that $\mu(xE) = \mu(E)$ for all $x \in G$ and for all $E \in \mathcal{B}$ and $\mu(G) = 1$.*

The proof can be found in the classic text on ergodic theory [Walters, 1982].

This unique measure μ is called the Haar measure. The Haar measure also satisfies that $\mu(Ex) = \mu(E), \forall x \in G$ and $\forall E \in \mathcal{B}$, because for each fixed point $x \in G$ the measure $\mu_x(E) = \mu(Ex)$ is invariant under rotations and therefore equal to μ.

---

[2]That is why I refer to the classic text for the proof instead of proving it myself.

If U is a non-empty open subset of G, then it has Haar measure different than zero.

This is because $G = \cup_{g \in G} gU = g_1 U \cup g_2 U \cup ... \cup g_k U$ for compactness.

Therefore if $\mu(U) = 0$ then

$$\mu(G) \leq \sum_i \mu(g_i U) = \sum \mu(U) = 0 \neq 1 \qquad (6.2)$$

which is a contradiction.

It is also easy to see that if G is infinite then the measure of all points is zero. In other words, the Haar measure does not contain atoms[3].

For the group of the circle $S^1 = \{z \in \mathbb{C} \mid\mid z \mid = 1\}$ the Haar measure is the same as the circular normalized Lebesgue measure. Moreover for the torus Tor^n, the Haar measure is the direct product of the Haar measure in S^1. Please keep in mind the definition of the torus based on the product of circles.

---

[3]As an interesting note aside, a non-integrable function is the Dirichlet function defined as $f : \mathbb{R} \rightarrow \mathbb{R}$ such that $f(x) = \begin{cases} 0, x \in \mathbb{Q} \\ 1, x \in \mathbb{I} \end{cases}$ which is topologically discontinuous at every point and any interval in the domain has measure equal to zero.

Definition 66. *Measure Preserving Transformation. In general if* $T : (X_1, \mathcal{B}_2, \mu_1) \to (X_2, \mathcal{B}_2, \mu_2)$, *we say that* T *preserves measure if the following conditions hold:*

1. T *is measurable, i.e.* $T^{-1}(A) \in \mathcal{B}_1$, *for all* $A \in \mathcal{B}$ *or* $T^{-1}(\mathcal{B}_2) \in \mathcal{B}_1$

2. $\mu_1(T^{-1}(A)) = \mu_2(A)$, *for all* $A \in \mathcal{B}_2$

The invariance of the measure with respect to the action of the group G means that for any set $A \in X$ and for any element $g \in G$ we have:

$$\mu(A) = \mu(T_g^{-1}(A)) \tag{6.3}$$

Now, I will give some definitions and theorems linked to integrable functions and their spaces. These will be necessary for the proofs of the fundamental theorems like the "Birkhoff Ergodic Theorem".

Definition 67. *Integrable Function. We say that a function* $f : X \to C$ *is integrable if*

$$\int_X | f | \, d\mu < \infty \tag{6.4}$$

Definition 68. *Space of Integrable Functions. We denote with* $L^1(X, \mathcal{B}, \mu)$ *the space of integrable functions* $f : X \to C$.

We say that the space $L^1(X, \mathcal{B}, \mu)$ or simply $L^1(\mu)$ is a Banach space with norm $\|f\|_1 = \int | f | \, d\mu$. Likewise, the space $L^1(X, \mathcal{B}, \mu)$ with

$p \geq 0$ denotes the space of all functions $f : X \to C$ such that $\mid f \mid^p$ is integrable and the formula $\|f\|_p = (\int \mid f \mid^p d\mu)^{\frac{1}{p}}$ defines a norm on $L^p(X, \mathcal{B}, \mu)$.

Observation. The space $L^2(\mu)$ is a Hilbert space.

Definition 69. *Separable Space 1. A Hilbert space is separable if it contains a dense countable set.*

Definition 70. *Separable Space 2. The space $L^2(X, \mathcal{B}, \mu)$ is separable if and only if $(X, \mathcal{B}, \mu)$ has a countable basis in the sense that there is a sequence of elements $|E_n|_1^n$ on $\mathcal{B}$ such that for all $\epsilon > 0$ and for all $B \in B$ with $\mu(B) < \infty$ there is some n with $\mu(B \triangle E_n) < \epsilon$ where BE_n is the symmetric difference between B and E_n.*

Theorem 4. *If X is a metric space, $\mathcal{B}$ is a Borel σ- algebra and μ is any probability measure on $(X, \mathcal{B})$ then $(X, \mathcal{B}, \mu)$ has a countable basis.*

The proof of this theorem can be found in the classic text [Rudin, 1976].

One of the concepts in which this work revolves around is that of "automorphism". This concept is closely related to dynamical systems, however it needs more structure because it is defined over a measurable space and as such, it requires that the function is measurable and not only measure-preserving.

Definition 71. *Automorphism. An automorphism on a measure space $(X, \mathcal{B})$ then $(X, \mathcal{B}, \mu)$ is an injective map T from the space X onto itself such that for all $A \in \mathcal{B}$ we have that $T(A)$ and $T^{-1}(A)$ are both $\in \mathcal{B}$ and*

$$\mu(A) = \mu(TA) = \mu(T^{-1}A) \tag{6.5}$$

In this case we have that T preserves measure.

The idea of automorphism is nothing but a function of a set onto itself; so every element of the set is mapped to one element of the same set (sometimes itself) and we normally associate an inverse to this and that is why groups are important in this case.

As a note aside, automorphisms are the way in which activation functions act in neural networks see [Thiede, et. al, 2021] for example of this.

Definition 72. *Simple Function. Let $A_1, A_2, A_3, ..., A_n \in \mathcal{B}$ pairwise disjoint and let $a_1, a_2, a_3, ..., a_n \in \mathbb{R}$ then we say that a function $f : X \to \mathbb{R}$ is simple and it can be expressed as follows:*

$$f(x) = \sum_{i=1}^{n} a_i \chi_A \tag{6.6}$$

where $a_i \in \mathbb{R}$ is the characteristic function of A or:

$$\chi_A(x) = \begin{cases} 0, \text{x} \notin A \\ 1, \text{x} \in A \end{cases} \tag{6.7}$$

An important relationship between characteristic functions and transformations of the base space onto itself is expressed in the following theorem:

Theorem 5. *Let A be an arbitrary subset of X and let $T : X \to X$ a function; then:*

$$\chi_A \circ T = \chi_T^{-1}(A) \tag{6.8}$$

Proof. Two function are equal if they have the same domain, the same codomain and they coincide on the evaluation of every point of the domain. So:

1. $Dom(\chi_A \circ T) = Dom(T) = X$

   $Dom(\chi_{T^{-1}(A)} = X$ due to the fact that $T^{-1} \subset X$

2. $Cod(\chi_A \circ T) = \{0, 1\}$

   $Dom(\chi_{T-1(A)} = \{0, 1\}$

3. Let $x \in X$ then we need to prove that $\chi_{T^{-1}(A)}(x) = \chi_A(T(x))$. For this, the proof will be divided in two parts:

Case 1. $x \in T^{-1}(A)$

Case 2. $x \notin T^{-1}(A)$

Now, $x \in T^{-1}(A)$ if and only if $T(x) \in A$ because $T^{-1}(A) = \{x \in X \mid T(x) \in A\}$, and if $x \notin T^{-1}(A)$ if and only if $T(x) \notin A$; therefore

For case 1. Suppose that $T(x) \in A$, that is $x \in T^{-1}(A)$

a) $\chi_A(T(x)) = 1$ because $x \in A$

b) $\chi_{T^{-1}(A)}(x) = 1$ because $x \in T^{-1}(A)$

For case 2. Suppose that $T(x) \notin A$ that is $x \notin T^{-1}(A)$

a) $\chi_A(T(x)) = 0$ because $x \notin A$

b) $\chi_{T^{-1}(A)}(x) = 0$ because $x \notin T^{-1}(A)$

$\blacksquare$

The elements of the proof can formalize even how we count. Counting in normal life is a) defining a set A, b) applying the composite characteristic function and c) integrating over it.

The result is the number of elements in the set. In this way, we also obtain the size of the set.

How would you "teach" a computer to count? Well, with the characteristic function and the theorem above. Now you see the power of mathematical foundations?

Theorem 6. *Let $f : X \to [0, \infty]$ a measurable function, then there is a family of simple functions s_n in X, for $n \in \mathbb{N}$ such that*

i) $0 \leq s_1 \leq s_2 \leq ... \leq f$

ii) $s_n^+ \to f(x)$ if $n \to \infty$ for all $x \in X$

The proof can also be found in the classic text [Rudin, 1976].

Theorem 7. *Lebesgue monotone convergence. Let $\{f_n\}$ a sequence of measurable functions on X and suppose that*

i) $0 \leq f_1(x) \leq f_2(x) \leq ... \leq \infty$

ii) $f_n)(x) \to f(x)$ if $n \to \infty$ for all $x \in X$

then f is measurable and

$$\int_X f_n d\mu \rightarrow \int_X f d\mu \qquad (6.9)$$

if $n \rightarrow \infty$.

This is a classic theorem and the proof can be found in [Rudin, 1974].

Lemma 1. *Fatou. If $f_n : X \rightarrow [0, \infty]$ is measurable for each positive integer n, then:*

$$\int_X (\lim_{n \to \infty} \inf f_n) d\mu \leq \lim_{n \to \infty} \inf \int_X f_n d\mu \qquad (6.10)$$

The proof can be found in [Rudin, 1974].

Theorem 8. *If $T : X \rightarrow X$ and preserves measure, then T defines a linear transformation T^* on L^p given by $T^*(f) = f \circ T$ on itself, such that $\mid T^*(f) \mid_p = \mid f \mid_p$. In other words, T^* preserves the metric i.e. T is isometric.*

Proof. What we need to prove is that given $\int_X \mid f \mid^p d\mu < \infty$ then $\int_X \mid f \circ T \mid^p d\mu < \infty$. First we will prove it for simple functions and then we will generalize for any $f \in F^p$.

Let f be a simple function, then

$$\int_X \mid f \mid^p d\mu = \int_X \mid \sum_i a_i \chi_{Ai} \mid^p d\mu = \int_X \sum_i \mid a_i \mid^p \chi_{Ai} d\mu = \quad (6.11)$$

$$\sum_i \mid a_i \mid^p \int_X \chi_{Ai} d\mu = \sum_i \mid a_i \mid^p \mu(A_i) \quad (6.12)$$

But we also know that

$$\sum_i a_i \chi_{Ai} \circ T = \sum_i a_i (\chi_{Ai} \circ T) = \sum a_i \chi_{T^{-1}(A_i)} \quad (6.13)$$

hence

$$\int_X \mid (\sum_i a_i \chi_{Ai}) \circ T \mid^p d\mu = \int_X \mid \sum_i a_i (\chi_{T^{-1}A_i}) \mid^p d\mu \quad (6.14)$$

$$= \int_X \sum_i \mid a_i \mid^p (\chi_{T^{-1}A_i}) d\mu = \sum_i \mid a_i \mid^p \mu(T^{-1}(A_i)) = \quad (6.15)$$

$$= \sum_i \mid a_i \mid^p \mu(A_i) \quad (6.16)$$

$$= \int_X \mid \sum_i a_i \chi_{A_i} \mid^p d\mu \quad (6.17)$$

therefore, if f is simple then

$$= \int_X \mid f \mid^p d\mu = \int_X \mid f \circ T \mid^p d\mu \qquad (6.18)$$

An application of the Lebesgue monotone convergence completes the proof for all $f \in L^p$.

∎

6.2   Characters

Another important tool for the development of this analysis and most importantly to understand the asymptotic behavior of dynamical systems is the concept of character. Specifically characters will be useful to analyze ergodic transformations on the circle especially because one point of high interest is the continuous homomorphisms on the circle group S^1.

Definition 73. *Character. Let G be an Abelian group. We denote by $\widehat{G}$ the collection of all continuous homomorphisms of G on the unit circle S^1. We say that the members of $\widehat{G}$ are the characters of G.*

It is relevant to note that under the multiplication of functions, $\widehat{G}$ is an Abelian group. With the compact open topology, $\widehat{G}$ becomes a commutative group locally compact.

If $G = S^1 = \{z \in \mathbb{C} \mid \mid z \mid = 1\}$ then each element of $\widehat{S^1}$ is of the form $z \to z^n$ for some $n \in \mathbb{Z}$. Moreover, the group of torus characters of dimension n or (Tor^n) is isomorphic to $(\mathbb{Z}^n, +)$ and each $\gamma \in \widehat{Tor^n}$ is of the form:

$$\gamma(z_1, z_2, ..., z_n) = z_1^{p_1}, z_2^{p_2}, ...z_n^{p_n} \tag{6.19}$$

for some $(p_1, p_2, ..., P_n) \in \mathbb{Z}^n$.

Given the above, we can withdraw the following two results:

1. $\widehat{G}$ has a countable topological basis if and only if G has a countable topological basis

2. $\widehat{G}$ is compact if and only if G is discrete

Combining these two results, G is compact with an associate metric (or a metric group) if and only if $\widehat{G}$ is a discrete countable group. This allows to transform some problems related to Abelian compact metric groups into problems of discrete countable Abelian groups. This will be clear when we analyze Torus's transformations.

Theorem 9. *Duality Theorem.* $\widehat{G}$ *is naturally isomorphic (as a topological group) to G; the isomorphism is given by the map $\alpha \to a$ where $\alpha(\gamma) = \gamma(\alpha)$ for all γ in G.*

The proof can be found in the influential text [Petersen, 1983].

If G_1 and G_2 are locally compact Abelian groups, then $\widehat{G_1 \times G_2} = \widehat{G_1} \times \widehat{G_2}$.

All the characters of $\widehat{G_1 \times G_2}$ are of the form $(x, y) \mapsto \gamma(x)\delta(y)$ where $\gamma \in G_1$ and $\delta \in G_2$.

Now, let G be compact, the members of $\widehat{G}$ are mutually orthogonal members of $L^2(\mu)$, where μ is the Haar measure.

Finally, if G is a compact group, G is metric if and only if G has a countable topological basis. It is important to note that a set is countable iff its cardinality (or the number of elements in it) is either finite or equal to the tiniest infinite set which we denote by $\aleph_0$ (Aleph-Zero). A set is denumerable iff its cardinality is exactly $\aleph_0$. A set is uncountable iff its cardinality is greater than $\aleph_0$. For more on this topic I can refer the reader to [Walters, 1982].

6.3   Torus Endomorphisms

As I have mentioned before, the torus can be seen in a multiplicative way as Tor^n or in an additive way as $\mathbb{R}^n/\mathbb{Z}^n$, where $\mathbb{R}^n$ is the additive group of the Euclidean n-dimensional space and $\mathbb{Z}^n$ is the subgroup of $\mathbb{R}^n$ consisting of all the points of integer coordinates.

In this section I will prove all theorems as the arguments used in them are very useful on problem solving strategies, like dividing in cases, and the use of previous models and fundamental concepts of probability like Borel sets.

Let us start with the fact that an isomorphism on topological groups from the n-torus Tor^n onto $\mathbb{R}^n/\mathbb{Z}^n$ is given by:

$$(e^{2\pi i x_1}, ..., e^{2\pi i x_n} \mapsto (x_1, ..., x_n) \tag{6.20}$$

Theorem 10. *The following properties hold:*

i) Each closed subgroup of S^1 is either S^1 itself or a finite cyclic group consisting of all the p roots of unity for an integer $p > 0$

ii) The only automorphisms of S^1 are the identity and the mapping $z \mapsto z^{-1}$

iii) The only homomorphisms of S^1 are the mapping $\phi_n(z) = z^n$ with $n \in \mathbb{Z}$

Proof. Let d be the usual Euclidean metric on S^1. Remember that this metric is an invariant rotation metric on S^1.

i) Let H be a closed subgroup of S^1. If H is infinite then it has a limit point, thus $\forall \epsilon > 0, \exists a, b \in H$ with $d(a, b) < \epsilon$ and $a \neq b$. Then $d(b^{-1}a, 1) < \epsilon$ and therefore the powers of $b^{-1}a$ are ϵ-dense on S^1.

Therefore H is ϵ-dense on S^1 and $H = S^1$.

If H is finite and has p elements, then $a^p = 1$, $\forall a \in H$. Thus, each elements of H is the p-th root of unity and given that there are p elements in H, H must consist of all the p-th roots of unity.

ii) Let $\theta : S^1 \to S^1$ an automorphism. We have that $\theta(1) = 1$. Given that -1 is the only element of S^1 of order 2, we have that $\theta(-1) = -1$. Given that i and $-i$ are the only elements of order 4, then either $\theta(i) = i$ and $\theta(-i) = -i$ or $\theta(i) = -i$ and $\theta(-i) = i$

Now consider the first case. Given that θ maps into intervals, the interval $\overrightarrow{[1, i]}$ is mapped into itself or into $\overrightarrow{[i, 1]}$. Bear in mind that all intervals are considered anti-clockwise. But given that $\overrightarrow{[1, i]}$ does not contain -1, then it cannot be mapped into $\overrightarrow{[i, 1]}$, hence $\theta\overrightarrow{[1, i]} = \overrightarrow{[1, i]}$.

The only element of order 8 in $\overrightarrow{[1, i]}$ is $e^{\frac{\pi i}{4}}$, hence this should be defined by θ. Therefore $\theta\overrightarrow{[1, e^{\pi i/4}]}$. By induction we can prove that $\theta(e^{\pi i/2^k}) = e^{2\pi i/2^k}$ for $k > 0$.

It follows that θ defines all the 2^kth roots of unity $\forall k > 0$ and then it is the identity. In the second case I will show that $\theta(e^{\pi i/2^k}) = e^{2\pi i/2^k}$ for $k > 0, \forall k > 0$ and then $\theta(z) = z^{-1}$, $z \in S^1$.

iii) Let : $\theta : S^1 \to S^1$ an endomorphism. If is not trivial, its image $\theta(S^1)$ is a closed connected subgroup of S^1, hence $\theta(S^1) = s^1$ by point i).

The kernel $Ker(\theta)$ is a closed subgroup of S^1, then either $Ker(\theta) = S^1$ or $Ker(\theta) = H_p$ which is the group of all the roots of unity for some p.

The first case corresponds to a trivial θ. If $Ker(\theta) = H_p$ let $\alpha_p : S^1/H_p \to S^1$ the isomorphism induced by θ; in other words, (not exactly words) $\theta_1(zH_p) = \theta(z)$.

Then $\theta\alpha_p^{-1}$ is an automorphism of S^1 and by the result in section ii) of this proof, either $\theta_1\alpha_p^{-1}(z) = z, \forall z \in S^1$ or $\theta_1\alpha^{-1}{}_p(z) = z^{-1}, \forall z \in S^1$. Thus either $\theta(z) = \theta_1(zH_p) = \theta_1\alpha_p^{-1}(z^p) = z^p, \forall z \in S^1$ or $\theta(z) = z^{-p}, \forall z \in S^1$.

$\blacksquare$

Theorem 11. *Each endomorphism $A : Tor^n \to Tor^n$ is of the form*

$$A(z_1, z_2, ...z_n) = \qquad (6.21)$$

$$= \left(z_1^{a_{11}} \cdot z_2^{a_{12}} \cdot ... \cdot z_n^{a_{1n}}, z_1^{a_{21}} \cdot z_2^{a_{22}} \cdot ... \cdot z_n^{a_{2n}}, z_1^{a_{n1}} \cdot z_2^{a_{n2}} \cdot ... \cdot z_n^{a_{nn}} \right) \qquad (6.22)$$

where $a_{ij} \in \mathbb{Z}$. Or in additive notation:

$$A\left(\begin{bmatrix} x_1 \\ \vdots \\ x_n \end{bmatrix} + \mathbb{Z}^n \right) = [a_{ij}] \begin{bmatrix} x_1 \\ \vdots \\ x_n \end{bmatrix} + \mathbb{Z}^n \qquad (6.23)$$

where $[a_{ij}]$ is the $n \times n$ matrix with a_{ij} on the (i, j)-th entry.

Proof. Let $\pi_i : Tor^n \to Tor^n$ the n-th coordinate projection. Then $\pi_i \circ A : Tor^n \to Tor$ is a homomorphism; and due to the fact that the only homomorphisms from the n-torus Tor^n onto S^1 are the mappings of the form:

$$(z_1, z_2, ...z_n) \mapsto z_1^{m_1} \cdot z_2^{m_2} \cdot .... \cdot z_n^{m_n} \tag{6.24}$$

then

$$\pi_i \circ A(z_1, z_2, ...z_n) = \left(z_1^{a_{i1}} \cdot z_2^{a_{i2}} \cdot .... \cdot z_n^{a_{in}}\right) \tag{6.25}$$

where $a_{ij} \in \mathbb{Z}$.

It is convenient to note that the endomorphism A maps the n-torus Tor^n onto itself if and only if $det\left[a_{ij}\right] \neq 0$.

Also that A is an automorphism of Tor^n if and only if $det\left[a_{ij}\right] =\pm 1$.

Therefore, the surjective endomorphisms of Tor^n are on a one-to-one correspondence with the matrices of integer entries whose determinant is not equal to zero.

Let $A : Tor^n Tor^n$ an endomorphism. Now, we must consider the mapping $\widehat{A} : \widehat{Tor^n} \to \widehat{Tor^n}$ acts as a function on $\mathbb{Z}^n$ when $\widehat{Tor^n}$ links to $\mathbb{Z}^n$ through the isomorphism:

$$\gamma \mapsto \begin{bmatrix} x_1 \\ \vdots \\ x_n \end{bmatrix} \tag{6.26}$$

when $\gamma(z_1, z_2, ..., z_n) = z_1^{m_1} \cdot z_2^{m_2} \cdot .... \cdot z_n^{m_n}$

Theorem 12. *Let G be a compact Abelian topological group and let $A : G \to G$ a surjective endomorphism; then A preserves Haar measure.*

Proof. Let $A : G \to G$ a continuous endomorphism and let μ be the Haar measure on G. We define a normalized measure of the Borel subsets of G as $\mu(E) = m(A^{-1}(E))$.

Now, $\mu(Ax \cdot E) = m(A^{-1}x \cdot E)) = m(x \cdot A^{-1}E) = \mu(E)$, given that A maps G into itself, we have that μ is an invariant rotation and therefore $\mu = m$. This is due to the unity property of the Haar measure. Therefore A preserves Haar measure. ∎

6.4   Ergodicity

The etymology of the word ergodic comes from two Greek words: $\acute{\epsilon}\rho\gamma o$ meaning *work* and $\breve{o}\delta\acute{o}c$ meaning *trail*[4]. The word as we know it in mathematics, was introduced by Ludwig Boltzmann while developing his constant (the Boltzmann constant or kB is the proportionality factor that relates the average relative kinetic energy of particles in a gas mole)[5].

When studying dynamical systems it is important to know -among other things- the asymptotic behavior of the many orbits, especially if at any moment the function passes through a given point or through the neighborhood of the given point again[6]. This is when for all $K \in \mathbb{N}$ there is a $k > K$[7] such that $T^k(p) \in E$, for $E \subset X$ and if it does, a good question is "how many times does it do it?". In this case, p is a recurrent point of E for a given $T : X \to X$.

---

[4]I like this last particle as it is used in Homer's Odyssey to describe the path or road built step by step by the Greek hero. This is another great analogy for the application of the *Plata-Boere third iteration recurrent theory (THIRTY)* as it is through step by step and adventure by adventure that Odysseus (or its Latin version *Ulysses*) built his story.

[5]In statistical mechanics one of the main objectives is to establish results over concepts studied with two particles in classical mechanics but with a lot of particles; in fact, the amount of particles is Avogadro's number which is 6.022×10^{23} number of particles in a mole. Boltzmann concluded in his hypothesis that for large systems of interacting particles in equilibrium, the time average along a single trajectory equals the space average.

[6]For a great example on asymptotic behavior and probabilities, see [Bisewski, 2020].

[7]This means a large enough number.

The fact that k always exists implies that T returns an infinite number of times to the set E excluding the case of finite recurrence. An interesting question is what would all the particles that recur to E form as orbits in a dynamical system with respect to E? Would they form the actual set E?

In many occasions we would like to predict the behavior of some systems, for example the weather in a definite region. And to do that we analyze the previous behavior of the system and its recurrent points; in other words we identify patterns.

The idea is that if it rains with certain intensity on any given day, we would be interested in knowing if it could rain again in the future with a similar intensity. This is what we call now "predictive analytics".

Now the reader can see that the *Plata-Boere third iteration recurrent theory (THIRTY)* is the cornerstone of predictive analytics, because it is precisely in those iterations when we establish patterns. In other words, it is in this process that noise becomes a signal.

The idea that something that happened in the past can happen again in the future is fundamental to many models in AI and pattern recognition and is closely related to the following theorem:

Theorem 13. *Poincaré Recurrence Theorem. For any compact automorphism $T : X \to X$ that preserves measure and for any Borel set $E \in \mathcal{B}$, we have that almost all points $x \in E$ are recurrent points of E.*

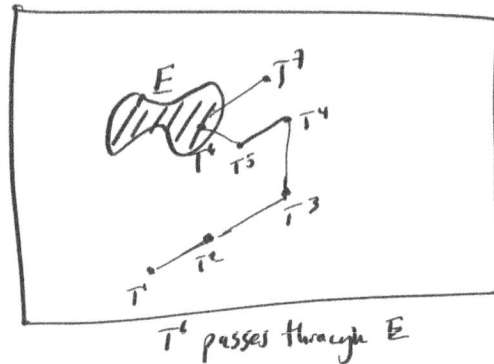

T^6 passes through E

Proof. Let N be a subset of E that consists of all points $x \in E$ that do not return to E. Also let $N \in \mathcal{B}$ such that

$$N = E \cap (\cup_{n=1}^{\infty} T^{-n}(X/E)) \tag{6.27}$$

If $x \in N$ then all the points of the form $T^n(x)$ with $n = 1, 2, 3, ...,$ do not belong to E and then $T^n(x) \notin N$. That is, $x \notin T^{-n}(N)$. Therefore $N \cap T^{-n}(N) = \emptyset$ with $n = 1, 2, 3, ...$

This implies that all the sets $N, T^{-1}(N), T^{-2}(N), ...$ are disjoint. As a matter of fact, for $0 < n_1 < n_2$ we have that

$$T^{-n_1}(N) \cap T^{-n_2}(N) = T^{-n_1}(N \cap T^{(-n_1-n_2)}(N)) = \emptyset \tag{6.28}$$

therefore

$$1 \geq \mu(\cup_{n=0}^{\infty} T^{-n}(N)) = \sum_{n=0}^{\infty} \mu(T^{-n}(N)) = \sum_{n=0}^{\infty} \mu(N) \qquad (6.29)$$

The last inequality holds only if $\mu(N) = 0$.

■

Another way to enunciate the recurrence theorem is:

Given $T : X \rightarrow X$ a transformation that preserves measure in a measure space $(X, \mathcal{B}, \mu)$ and $E \in \mathcal{B}$ with positive measure, then almost all points of E return to E an infinite number of times under positive iterations of T.

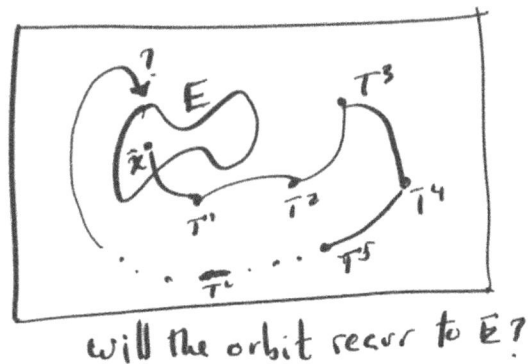

will the orbit recur to E?

Definition 74. *Temporal Mean or Time Mean. If the following limit exists:*

$$\dot{f} = \lim_{n \to \infty} \frac{1}{n} \sum_{k=0}^{n-1} f(T^k(x)) \tag{6.30}$$

for a function $f \in L^1$ with $x \in X$ and $n \in N$, then $\dot{f}$ is called the time mean or temporal mean in a dynamical system generated by T.

We also say that this mean is the mean on all the trajectory. Remember that the trajectory of a point x is the set of all the points of the form $T^k(x)$.

For example $\frac{1}{n} \sum_{k=0}^{n-1} \chi_E(T(x))$, where χ_E is the characteristic function of E, is the average number of times the trajectory of point x passes through the set E from time $t = 0$ up to $t = n - 1$.

Definition 75. *Spatial Mean or Space Mean. If the following limit exists:*

$$\bar{f} = \int_X f(x)d\mu \tag{6.31}$$

for a function $f \in L^1$ then $\bar{f}$ is called the spatial mean or space mean of a given dynamical system generated by T.

In a dynamical system, we are interested in making statistical predictions about the orbits of T. A good way to do this is counting the orbits or some orbit that passes through some part of the space; and if it does, then verify the frequency with which it does it.

In the case of the circle rotations, we can measure how many times a function T passes through a given interval. In other words, we would like to calculate how many times T^i passes through a set E after n iterations, and moreover get a statistical parameter out of it (like the average):

$$\frac{\#(i, \ni T^i(x) \in E)}{n}, i = 1, ..., n \tag{6.32}$$

This expression represents the relative frequency of the number of times that a transformation $T^i(x)$ passes through a set E.

Bear in mind that relative frequency is the classic approach to probability where the probability of the occurrence of an event denoted by $P(x)$ for a given event x is given by the total number of positive occurrences divided by the total number of occurrences.

Another relevant question is what happens when the number of iterations is really large, i.e. when $n \to \infty$[8]:

$$\lim_{n \to \infty} \frac{\#(i, \ni T^i(x) \in E)}{n} \tag{6.33}$$

---

[8]This is what we call the asymptotic behavior of the system.

In the example of the circle rotations, suppose that the rotation T is $\frac{\pi}{2}$, then the statistical question is how many times $T^i(x)$ passes through the interval $(\frac{-\pi}{4}, \frac{\pi}{4})$ after a number of iterations and if so, with what relative frequency.

Suppose now that we want to know how many times T passes through that interval after 4 iterations, starting at the point $\theta = 0$ (for $e^{2\pi i \theta}$), intuitively we can infer that T passes through the mentioned interval every 4 times. Then the relative frequency is 1 in 4:

$$\frac{\#(i, \ni T^i(x) \in (\frac{-\pi}{4}, \frac{\pi}{4}))}{n} = \frac{1}{4} \tag{6.34}$$

and

$$\lim_{n \to \infty} \frac{\#(i, \ni T^i(x) \in (\frac{-\pi}{4}, \frac{\pi}{4}))}{n} = \frac{\#\{\frac{-\pi}{4}, \frac{\pi}{4})) \cap \mathcal{O}(x)\}}{n} \tag{6.35}$$

where $\mathcal{O}(x)$ is the orbit pf x under T.

Now, if the rotation does not have periodic points, which by the way is the most important case for the ergodic analysis, then we would tend to think that the number of times T^i passes through the mentioned interval is precisely the length of the interval (or the its measure) as seen in the previous section on the rotation number theorem.

Now the reader will notice that we are defining measures with iterations, which is along with the spirit of the book.

The *Plata-Boere third iteration recurrent theory (THIRTY)* applies also to this case as one or two iterations would not be enough to define a measure, but enough number of iterations will. And there is a point "the third iteration" in which quantitative has a qualitative effect. That iteration is when we start seeing the pattern.

Enough iterations combined with the characteristic function will define a measure, a probability and even a prediction.

This is mainly due to the fact that at each "turn" that T does over the circle S^1, does not land on the same point as the previous round; in other words, it does not hold that $T^n(x) = x =$, then the rotation always lands on different points as it iterates more and more "filling up" the interval when $n \to \infty$.

And precisely Birkhoff Ergodic Theorem states that:

$$\lim_{n\to\infty} \frac{\#(i, \ni T^i(x) \in E)}{n} = \mu(E) \tag{6.36}$$

6.5   Birkhoff Ergodic Theorem

While Poincaré states that there is a recurrence, Birkhoff states the measure of the recurrence. This is precisely how we define probabilities via iterations: a dynamical system (more precisely an ergodic dynamical system) iterating enough to recur and iterating enough to establish a measure.

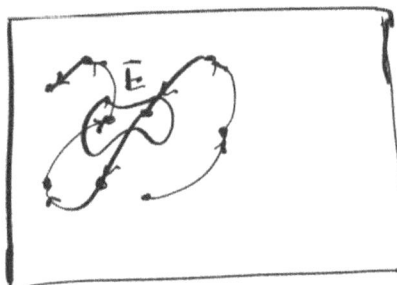

The measure of recurrence =
The probability of
passing through $\bar{E}$ =
AREA OF E !

Theorem 14. *Birkhoff Ergodic Theorem*[9]. *Let T be a transformation that preserves measure in a measure space X and let $f \in L^1(\mu)$. Then the ergodic average $\frac{1}{n}\sum_{k=0}^{n-1} f(T^k(x))$ converges for almost all x into a limit function $f^*(x)$ which is also contained in $L^1(\mu)$. The function f^* is constant on orbits i.e. $f^*(T^*(x)) = f^*(x)$ for almost all x. In the case of $\mu(X) < \infty$ we also have that $\int_X f d\mu = \int_X f^* d\mu$.*

To prove this theorem we will need the following theorem and although some authors know it as a lemma, I prefer to name it the "maximal ergodic theorem" which specifies the total sum of the measured values of the dynamical system and it is important to note that the initial value is zero.

---

[9]Also known as the Birkhoff-Khinchin Theorem.

Theorem 15. *Maximal Ergodic Theorem. Suppose that*

$$S_n(x) = \sum_{k=0}^{n-1} f(T^k(x)) \tag{6.37}$$

where $T^0(x) = 0$, for all x.

Birkhoff Ergodic Theorem is equivalent to prove that $\frac{1}{n}T^n(x)$ converges almost everywhere to a limit function $f^*(x)$.

Lemma 2. *Let $A = x \mid sup_{n\geq 0}s_n(x) > 0$ then $\int_A f(x)d\mu \geq 0$*

Proof. We have that:

$$max_{0\leq k\leq n}s_k(T(x)) = max\{0, f(T(x)), ..., f(T(x)) + ... + f(T^n(x))\} \tag{6.38}$$

$$max\{f(x), f(x) + f(T(x)), ..., f(T(x)) + f(T(x))... + f(T^n(x)) - f(x)\} \tag{6.39}$$

therefore

$$f(x) = max_{0\leq k\leq n+1}s_k(x) - max_{0\leq k\leq n}s_k(T(x)) = \varphi_{n+1}^*(x) - \varphi_n(T(x)) \tag{6.40}$$

where

$$\varphi_n^*(x) = max\{s_1(x), ..., s_n(x)\}, \varphi_n(x) = max\{0, s_1(x), ..., s_n(x)\}$$
(6.41)

Suppose now that $A_n = \{x \in X \mid \varphi_n(x) > 0\}$. Then, given that $\varphi_{n+1}^*(x) \geq \varphi_n^*(x)$, we have that

$$\int_{A_n} f(x)d\mu \geq \int_{A_n} \varphi_n^*(x)d\mu - \int_{A_n} \varphi_n(T(x))d\mu$$
(6.42)

Now, for $x \in A_n$, we have that $\varphi_n^*(x) = \varphi_n(x)$ and for $x \notin A_n$ we have that $\varphi_n(x) = 0$, thus,

$$\int_{A_n} \varphi_n^*(x)d\mu = \int_{A_n} \varphi_n(x)d\mu = \int_X \varphi_n(x)d\mu = \int_X \varphi_n(T(x))d\mu$$
(6.43)

And given that we have the measure preserving property on T, we have that

$$\int_{A_n} f(x)d\mu \geq \int_X \varphi_n(T(x))d\mu - \int_{A_n} \varphi_n(T(x))d\mu$$
(6.44)

$$= \int_{X-A_n} \varphi_n(T(x))d\mu \geq 0$$
(6.45)

All of the above given that $\varphi \geq 0$. Now, $A_n \subseteq A_{n+1}$ and $A = \cup_{n=1}^{\infty} A_n$. Hence

$$= \int_A f(x)d\mu = \lim_{n \to \infty} \int_{A_n} f(x)d\mu \geq 0 \qquad (6.46)$$

∎

6.6   Birkhoff Ergodic Theorem Proof

I wanted to highlight this proof as an entire subsection as it is of the utmost importance and the arguments needed to do it are also very useful while analyzing statistical behavior and other parameters of systems and processes.

I would like to add that random walks of chaotic dynamical systems in topological compact spaces that are recurrent in the sense of Poincaré are the foundations of the Monte Carlo method proposed originally by John Von Neumann and by extension recurrent random process are the basis of algorithms like random forest, Brownian motion and Markov chains.

Proof. For any two rational numbers a, b with $a < b$, we have that:

$$E_{a,b} = \{x \in X \mid \lim_{n \to \infty} inf \frac{1}{n} s_n(x) < a < b < \lim_{n \to \infty} sup \frac{1}{n} s_n(x)\} \quad (6.47)$$

We also know that $E_{a,b} \in \mathcal{B}$ and T is invariant.

In order to prove the existence of the limit $\lim_{n \to \infty} \frac{1}{n} s_n(x)$ almost everywhere, it is enough to show that $\mu(E_{a,b}) = 0$ for all a, b.

Now let us fix a and b and take $E = E_{a,b}$. Now consider the function:

$$g(x) = \begin{cases} \text{f(x)-b,x} \in E \\ 0, \text{x} \notin E \end{cases} \tag{6.48}$$

and applying the maximal ergodic theorem to this function, we obtain:

$$\int_{A_{(g)}} g(x) d\mu \geq 0 \tag{6.49}$$

where

$$A(g) = \{x \in X \mid sup_{n>1} \frac{1}{n} s_n(x, g) > 0\} \tag{6.50}$$

$$= \{x \in X \mid sup_{n>1} \frac{1}{n} s_n(x, f) > b\} \tag{6.51}$$

$E \subseteq A(g)$. Given that E is invariant and g disappears outside E, we have that $s_n(x, g) = 0$ for $x \in X$ i.e. $A(g) \subseteq E$. Therefore $A(g) = E$, and then expression (6.42) can be rewritten as follows, where the reader can appreciate the use of a measure to compare the size of a set.

$$\int_E f(x)d\mu \geq b\mu(E) \tag{6.52}$$

In an analogous way, consider the function:

$$g'(x) = \begin{cases} a - f(x), & \text{x} \in E \\ 0, & \text{x} \notin E \end{cases} \tag{6.53}$$

then

$$A(g') = \{x \in X \mid sup_{n>1}\frac{1}{n}s_n(x, g') > 0\} \tag{6.54}$$

$$= \{x \in X \mid inf_{n>1}\frac{1}{n}s_n(x, f) < a\} \tag{6.55}$$

Once again we have that $A(g') = E$ for the same reasons exposed above, and

$$\int_E f(x)d\mu \leq a\mu(E) \tag{6.56}$$

From (6.50) and (6.54) we have that $\mu(E) = 0$ given some a and b. Then the ergodic average converges almost everywhere to a limit function denoted by f^*, which will come clear after applying Fatou's Lemma.

$$\int_X |\frac{1}{n} \sum_{k=0}^{n-1} f(T^k(x))| \, d\mu \leq \frac{1}{n} \sum_{k=0}^{n-1} \int_X |f(T^k(x))| \, d\mu = \qquad (6.57)$$

$$= \int_X |f(x)| \, d\mu \qquad (6.58)$$

And because of Fatou's Lemma, we have that

$$\int_X |f^*| \, d\mu \leq \lim_{n \to \infty} \inf \int_X |\frac{1}{n} \sum_{k=0}^{n-1} f(T^k(x))| \, d\mu \leq \int_X |f(x)| \, d\mu$$
$$(6.59)$$

In this way we know that f^* is in $L^1(\mu)$.

Now suppose that $\mu(X) < \infty$. if f is a bounded function then the functions $\frac{1}{n} s_n(x, f)$ are also bounded. In fact they are bounded by the same constant that bounds f. Now, by the *convergence theorem* we have that

$$\int_X f^* d\mu = \lim_{n \to \infty} \int_X \frac{1}{n} s_n(x, f) d\mu \qquad (6.60)$$

$$= \lim_{n \to \infty} \int_X \frac{1}{n} \sum_{k=0}^{n-1} f(T^k(x)) d\mu \qquad (6.61)$$

$$= \int_X f d\mu \tag{6.62}$$

The application of this concept in data science is clear when working with penalized regressions. In this case we often encounter two popular ones: *lasso* and *ridge* regressions. The first one uses an L^1 penalty and the second one uses an L^2 penalty type. This refers precisely to what we are seeing in this theorem: when we penalize *lasso*, we do it on the one-dimensional norm $L^1(\mu)$; in other words, a measure in an L^1 space. However, when we work on the *ridge* case we do it in an L^2 space.

The first one applies the penalty on the sum of its absolute values (remember that the absolute value is also the measure of an interval), while the second one applies it on the Euclidean metric (remember the Euclidean metric is the length in a two-dimensional space).

Moreover, the popular *elastic net regularization* penalizes using a combination of both, L^1 and L^2 norms of the corresponding parameter vectors in an L^1 and L^2 space respectively. Also remember that an L^2 space is also called a *Hilbert Space*.

Now, if f is not bounded but is contained in $L^1(\mu)$, then we can find a sequence $(f_k)_{k=1}^{\infty}$ of bounded functions that converge to f in the form of L^1.

Now, this looks like this:

$$\|f^* - \frac{1}{n}s_n(x, f)\|_1 \tag{6.63}$$

$$\leq \|f^* - f_K^*\|_1 f_k^* - \frac{1}{n}s_n(x, f)\|_1 + \|\frac{1}{n}s_n(x, f_k) - \frac{1}{n}s_n(x, f)\|_1 \tag{6.64}$$

$$\leq \|(f - f_k)\|_1 + \|f_k^* - \frac{1}{n}s_n(x, f_k)\|_1 + \frac{1}{n}\sum \|f_k - \|_1 \tag{6.65}$$

It follows that $\frac{1}{n}s_n(x, f) \to f^* \in \|L^1\|$ thus,

$$\int \frac{1}{n}s_n(x, f)d\mu \int_X f^* d\mu \tag{6.66}$$

But $\int_X \frac{1}{n}s_n(x, f)d\mu = \int_X f d\mu$ therefore

$$\int_X f^* d\mu = \int_X f d\mu \tag{6.67}$$

∎

Definition 76. *Ergodic Dynamical System. An ergodic dynamical system happens when the time measure equals the space measure.*

This definition is directly related to the statistical mechanics concept explained above. An example of this is if the rotations of $T : S^1 \to S^1$ from the circle onto the circle do not have periodic points, i.e. an irrational rotation, then there is an invariant measure (in this case it is the Lebesgue measure) for the intervals by which T passes through.

In the same way, if the rotation is rational, then the ergodic measure is invariant like the relative frequency in probability, being the total measure of the circle S^1: $\mu(S^1) = 1$.

In the following section we will analyze what happens with the eigenvalues of the linear transformations T and whether their induced operators -that are contained in the circle- are ergodic measures or not or even if the functions themselves are ergodic.

Eigenvalues are a great tool to classify and analyze such functions. Now I will approach ergodic transformations with non-decomposable functions.

Definition 77. *Decomposable Transformation. Let $T : X \to X$ be a transformation that preserves measure. If X is the union of two disjoint sets E and F of positive measure, and E and F are invariant under T, then the study of any property of T in X can be reduced to the study of the corresponding properties of T in E and T in F. If this is the case, then we say that T is decomposable.*

The most meaningful functions for this study are the non-decomposable. Such functions are called ergodic functions.

Ergodicity is one of the natural requirements for a transformation to "mix well" all the points in the space in which it acts.

Also the reader shout bear in mind that ergodic transformations act in topological compact spaces.

To give more precise example of ergodic transformations one has to reformulate with more precision the concept of ergodicity.

Definition 78. *Ergodic Transformation 1. We say that T is ergodic if and only if T has invariant trivial sets only; i.e. if and only if $\mu(E) = 0$ or $\mu(E - X) = 0$, as long as E is a measurable invariant set under T.*

In other words, if T is such that $T(E) = E$, then $\mu(E) = 0$ or $\mu(E - X) = 0$.

Now, another way to express the concept of an ergodic transformation is the following:

Definition 79. *Ergodic Transformation 2. Let $(X, \mathcal{B}, \mu)$ a probability space; i.e. a normalized measure space: $\mu(X) = 1$. A transformation T that preserves measure on $(X, \mathcal{B}, \mu)$ is ergodic if the only members of $E \in \mathcal{B}$ with $T(E) = E$ satisfy that $\mu(E) = 0$ or $\mu(E) = 1$.*

Definition 80. *Invariant Function. A function $g : X \to \mathbb{C}$ is invariant with respect to an automorphism T if:*

$$g(T^{-1}(x)) = g(T(x)) = g(x) \qquad (6.68)$$

This means that if we apply g to any point on the orbit, this will be lifted to a constant height on all the trajectory.

Lemma 3. *If the measure space X is a topological space with a numerable basis such that each non empty set has positive measure, and if T is an ergodic transformation that preserves measure on X, then for almost all $x \in X$ the orbit of x is dense everywhere.*

Proof. The orbit of x is not dense if and only if there is a non-empty open set G such that x is in the intersection of all $X - T^n(G)$. Given that this intersection is a disjoint invariant set of G and given that $\mu(G) > 0$, then the orbit has measure zero. If x does not belong to any denumerable class of measure zero sets, then x has a dense orbit.

■

Suppose that T is a rotation, i.e. $T(x) = cx$ in an Abelian compact group X with a numerable basis. If T is ergodic then by the previous lemma, there is at least one point, say x_0 whose orbit is dense.

Given that the transformation that sends x into x_0^{-1} is a homeomorphism, then it sends the orbit of $x_0(\{c^n x_0\})$ in a dense sequence.

However, the reader will notice that the image of the orbit consists exactly of the powers of c.

If F is an arbitrary function of X, another function g of X is defined by $g(x) = f(T(x))$. If we write $g = Uf$ then U is a mapping that operates over functions. This mapping U has important properties: i) U is linear i.e. $U(f + g) = Uf + Ug$ and $U(\lambda f) = \lambda Uf$ and ii) if T preserves measure, then U sends L^1 to itself becoming an isometry of L^1.

Suppose the opposite: that $\{c^n\}$ is dense. If f is the character of X (a continuous homomorphism on the circle group), then $f(cx) = f(c)f(x)$, thus f is an eigenvector of the unitary operator induced by T.

Given that the characters constitute an orthonormal complete set in L^2, each invariant function on L^2 can be expanded in terms of these characters.

Given that for unitary operators, eigenvectors with different eigenvalues are orthogonal, each invariant function on L^2 (each eigenvector with eigenvalue 1) is a linear combination of characters with eigenvalue 1.

As a matter of fact, if f is a character and $f(cx) = f(x)$ almost everywhere, then, by continuity, $f(cx) = f(x)$ everywhere and therefore $f(c^n x) = f(x)$ everywhere. The result follows if we make $x = 1$.

A rotation on the torus $T((x, y)) = ((bx, cy))$ is ergodic if and only if the multiplier coordinates (numbers a and b) are integrably independent, i.e. $b^n c^m = 1$ for integers m and n, implies that $n = m = 0$.

Theorem 16. *Equivalence Theorem. If $(X, \mathcal{B}, \mu)$ is a probability space and $T : X \to X$ preserves measure, then the following points are equivalent:*

1. *T is ergodic*

2. *As long as T is measurable and $(f \circ T)(x) = f(x), \forall x \in X$ then f is constant almost everywhere*

3. *As long as T is measurable and $(f \circ T)(x) = f(x)$ almost everywhere, then f is constant almost everywhere*

4. *As long as $f \in L^2(\mu)$ and $(f \circ T)(x) = f(x), \forall x \in X$ then f is constant almost everywhere*

5. *As long as $f \in L^2((\mu)$ and $(f \circ T)(x) = f(x)$ almost everywhere, then F is constant almost everywhere*

The proof can be found in the classic text [Walters, 1982].

Theorem 17. *The rotation $T(z) = az$ on the unit circle S^1 is ergodic (with respect to the Haar measure) if and only if a is not a root of unity.*

Proof. Suppose that a is a root of unity, then $a^p = 1$ for some $p \neq 0$. Let $f(z) = z^p$. Then $f \circ T = f$ and f is not constant almost everywhere. Then T is not ergodic because of point 2) of the equivalence theorem.

On the other hand, suppose that a is not a root of unity and that $f \circ T = f, f \in L^2(\mu)$. Let $f(z) \sum_{n \to -\infty}^{\infty} b_n \mathbb{Z}^n$ its Fourier series. Then $f(az) = \sum_{n \to -\infty}^{\infty} b_n a^n \mathbb{Z}^n$, hence $b_n(a^n - 1) = 0$ for each n. If $n \neq 0$ then $b_n = 0$. Therefore f is constant almost everywhere. And due to point 5) of the equivalence theorem, T is ergodic.

■

Theorem 18. *If G is a compact Abelian group with Haar measure, and $A : G \to G$ an continuous surjective endomorphism of G, then A is ergodic if and only if the trivial character $\gamma \equiv 1$ is the only $\gamma \in G$ that satisfies that $\gamma \circ A^n = \gamma$ for some $n > 0$.*

Proof. Suppose that as long as $\gamma A^n = \gamma$ for some $n \geq 1$ we have that $\gamma \equiv 1$. Let $f \circ A = f$ with $f \in L^2(\mu)$.

Suppose that $f(x)$ has a Fourier series $\sum a_n \gamma_n$ where $\gamma_n \in G$ and $\sum | a_n |^2 < \infty$. Then $_n\gamma_n(Ax) = \sum a_n \gamma_n(x)$, thus $\gamma_n, \gamma_n \circ A, \gamma_n \circ A^2, ...,$ are all different but their coefficients are equal and therefore zero.

Hence, if $a_n \neq 0, _n a^p) = \gamma_n$ for some $p \neq 0$, then $\gamma \equiv 1$ by assumption and hence f is constant almost everywhere. Therefore A is ergodic due to point 5) of the equivalence theorem.

Contrary to this, let A be ergodic and $\gamma A^n = \gamma, n > 0$. If n is the least integer, then $f = \gamma + \gamma A, + ... + \gamma A^{n-1}$ is invariant under A and not constant almost everywhere (being the sum of orthogonal functions), contradicting point 5) of the equivalence theorem.

■

We are specially interested in the case of the n-dimensional torus Tor^n. As we saw in the previous section (torus endomorphism), a surjective endomorphism $A : Tor^n \to Tor^n$ is given by the $n \times n$ matrix $[A]$ of integer entries and that $\widehat{Tor^n}$ can be identified with $\mathbb{Z}^n$ and the induced action $\widehat{A} : Tor^n \to Tor^n$ corresponds to the action of the transposed matrix $[A]_t$ in $\mathbb{Z}\mathbb{A}^n$.

Corollary. Let $A : Tor^n \to Tor^n$ be a continuous surjective endomorphism on the n-torus. Then A is ergodic if and only if the matrix $[A]$ does not have roots of unity as eigenvalues.

Proof. If A is not ergodic, the previous theorem on the existence of $q \in \mathbb{Z}^n, q \neq 0$ and $k > 0$ with $[A]_t^k q = q$ leads to the fact that $[A]_t^k$ has eigenvalue equal to 1; thus $[A]_t$ and therefore $[A]$ have the same kth root of unity as eigenvalue.

Contrary to this, if $[A]_t$ has the same kth root of unity as eigenvalue, then $[A]_t^k$ has eigenvalue 1. Therefore $([A]_t^k - I)(y) = 0$ for $y \in \mathbb{R}^n$, and given that the matrix $[A]_t$ has integer entries we can find $y \in \mathbb{Z}^n$. Therefore $[A]_t^k y = y$ and A is not ergodic by the result of theorem 15 (Maximal Ergodic Theorem).

■

Theorem 19. *If $T(x) = a \cdot A(x)$ is an affine transformation of the compact metric connected Abelian group G, the following are equivalent:*

1. *T is ergodic (relative to Haar measure)*

2. *Whenever $\gamma \circ A^k = \gamma$ for $k > 0$, then $\gamma \circ A = \gamma$ and*

3. *The smallest closed subgroup containing a and BG (where $Bx = x^{-1} \cdot A(x)$) is G in other words $[aBG] = G$*

4. *$\exists x_0 \in G$ with $\{T^n(x_0) : n \geq 0\}$ dense in G*

5. *$\mu(\{x : \{T^n x : n \geq 0 \text{ is dense }\}\}) = 1$*

The proof is part of the classic theory and can be found in [Walters, 1982].

This chapter connects the iterative nature of dynamical systems, the formalization and heart of probability and statistics and measure theory concepts, all of them under the algebraic structures of groups, and topological spaces. I recommend the reader to review the outstanding Berlanga-Epstein Theorem which specializes in mapping topological sigma-compact manifolds.

The real application of these concepts is when we analyze the behavior of neural networks with respect to their aggregation and activation functions, and possibilities to converge.

From the practical point of view, we can start a trial-and-error method to tune our neural-network-models, but this would take a long time compared to the laser focus solution that the analysis can give.

For example by understating equivalences in the models one can properly model accordingly and not just give shots in the dark with trial and error methods.

Once again, these tools will help understand the core and essence of machine learning algorithms and other statistical tools in the AI practice.

Chapter 7

Spectral Theory: An Introduction to Eigenvalues

> Everything we call real is made of things that cannot be regarded as real.
>
> *Philosophical Writings*
> *Niels Bohr*

Spectral theory tackles the geometric problem of some linear transformations $T : V \to V$, which is to find the direction in which vectors are stretched or shrunk by the transformation and its stretching factor too, and analyze how it transforms the geometry of surfaces. In fact spectral theory will be foundational to classify these geometric surfaces. As an introduction to spectral theory I will solve the initial problem with which the theory started.

The name "spectral theory" was introduced by David Hilbert while working on principal axis theorem[1].

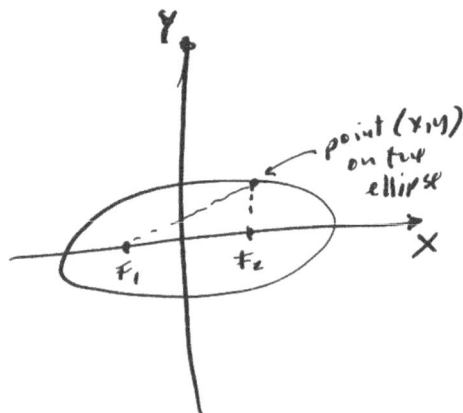

From the drawing above F_1 is one of the focus and F_2 is the other one. Also note from the drawing that the axes on the ellipse (the major and minor) are parallel to the coordinate axes and they are perpendicular to each other.

A typical problem in basic analytic geometry when studying the conics is to find the length semi major and semi minor axes of an ellipse and its orientation i.e. if it is elongated towards the X axis or the Y axis.

---

[1]Hilbert was actually working on the development of what we know now are "Hilbert spaces", which was expressed in terms of quadratic forms. The original spectral theorem was constructed as a version of the theorem on principal axes (i.e. completing the square) of an ellipse and ellipsoid.

The typical equation of an ellipse is parallel to the canonical Cartesian coordinate axes. The standard form equation is

$$\frac{(x-h)^2}{a^2} + \frac{(y-k)^2}{b^2} = 1 \qquad (7.1)$$

where if $a > b$ then a is the length of the semi major axis and b is the length of the semi minor axis and (h, k) is the center of the ellipse[2]

Another typical problem in basic analytic geometry is to convert an equation in its general quadratic form into the standard form to identify the elements of the conic curve i.e. its center, radius in the case of the circle, semi axes, orientation, directrix in the case of a parabola, foci, asymptotes in the case of a hyperbola, eccentricity, etc.

To do this, we normally use the method of "completing the square", which is generally taught in middle school. One important thing to note is that this method is actually the principal axis theorem and it is the geometric counterpart of the spectral theorem as stated by David Hilbert at the beginning of the 20th century.

The spectral theorem states that a linear transformation (represented by a matrix) can be diagonalized.

A diagonal matrix has a number of advantages! Firstly, if we can express a matrix A with an equivalent diagonal (see change of basis theorem

---

[2]The definition of an ellipse is: "the set of points on the two-dimensional plain $\mathbb{R}^2$ such that the sum of the distances between any point on the ellipse and the foci is constant".

in linear algebra), then we could eliminate crossed terms in quadratic equations[3].

Moreover, the spectral theorem leads to a canonical decomposition, called the spectral decomposition of a vector space.

In the end eigenvectors will allow us to express the same equations in a different coordinate axes. Conveniently these axes will be parallel to the canonical Cartesian ones. In basic linear algebra courses we learn that any two-dimensional vector $v \in \mathbb{R}^2$ can be expressed as a linear combination of the basis vectors. The canonical basis is $\hat{i} = (1, 0)$ and $\hat{j} = (0, 1)$. But we could choose any other basis as long as the vectors of the basis are linearly independent.

But let us solve the original problem that Hilbert was trying to solve (and although he did it for ellipsoids, we will start with ellipses) so we get a grasp of how the eigenvalues and eigenvectors started and understand their power, how the spectral decomposition works and hence we are able to interpret and use them in practical problems.

The ellipse standard equation can take a general quadratic form of $ax^2 + by^2 + cx + dy + exy + f = 0$ where a, b, c, d, e, f are constants in $\mathbb{R}$. Notice that in the general quadratic form we have a cross-term with x and y multiplying. All conics can be expressed as quadratic equations.

---

[3]A square matrix is said to be diagonalizable if it is similar to a diagonal matrix. That is, A is diagonalizable if there is an invertible matrix Q and a diagonal matrix D such that $A = QDQ^{-1}$. See also *dynamic similarity* in previous section.

In particular this could represent the equation of an ellipse whose axes are not parallel to the canonical axes in a Cartesian space. If the coefficient of the xy term is zero, then the ellipse's axes would be parallel to the canonical ones.

So a simple problem would be to identify the following equation of a rotated ellipse that looks more or less like this:

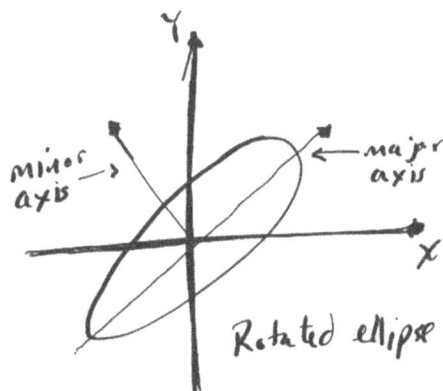

The equation is $5x^2 - 4xy + 5y^2 = 1$. The reader might appreciate the problem of the crossed term and the idea is to rotate it without deforming it so the major and minor axes are parallel to the canonical axes X and Y. In that way we could apply the classic analytic geometry concepts and get all the elements of the ellipse easily.

In this way, we could simply apply our basic knowledge on conics and perfectly describe the ellipse at hand even when it is rotated.

In matrix form,the above quadratic equation can be expressed as:

$$[x, y] \begin{bmatrix} 5 & -2 \\ -2 & 5 \end{bmatrix} \begin{bmatrix} x \\ y \end{bmatrix} = 1$$

I will not go into details, but the coefficient of the cross term is always divided by 2. This is due to a property of the ellipse. The reader might solve the above matrix multiplication to verify that the equation above is the same as the previous one.

It is important to note that the matrix $\begin{bmatrix} 5 & -2 \\ -2 & 5 \end{bmatrix}$ which I will call A is the matrix that generates the ellipse. And if we apply that transformation to any vector it will naturally "deform" the shape mapping vectors in different directions.

For example $(1, 1) \mapsto (1, 3)$ and $(3, 2) \mapsto (5, 4)$. Now the problem is to find the direction of the axes so we can transform the equation (rotate it) and eliminate the pesky crossed term. So how do we find this? Well, we need to find the scalars for which any vector (x, y) transformed by the matrix that generates the ellipse stays in the same direction. In other words a parallel vector.

The idea is that to find a scalar for which any vector under the transformation A will be a scalar multiple of itself so it does not "deform" it. In equation form we need to find a scalar λ such that:

$$A\bar{x} = \lambda\bar{x} \qquad (7.3)$$

We call λ the eigenvalue and in this way we can find the associated vector to that value, which will indicate the direction of the transformation.

To find the value(s) we need to manipulate with basic linear algebra operations the following:

$$\begin{bmatrix} 5 & -2 \\ -2 & 5 \end{bmatrix} \begin{bmatrix} x \\ y \end{bmatrix} = \lambda \begin{bmatrix} x \\ y \end{bmatrix}$$

And moving all terms on the right hand side of the equation to the left we have:

$$\begin{bmatrix} 5 - \lambda & -2 \\ -2 & 5 - \lambda \end{bmatrix} = 0$$

And to solve for lambda we calculate the determinant of the matrix above obtaining: $(5 - \lambda)^2 - 4 = 0$. Hence $\lambda = 7$ and $\lambda = 3$.

If we substitute the two values of lambda in:

$$\begin{bmatrix} 5 - \lambda & -2 \\ -2 & 5 - \lambda \end{bmatrix} \begin{bmatrix} x \\ y \end{bmatrix}$$

we have that for $\lambda = 7$ we get $y = -x$, so the eigenvector is of the form $(-1, 1)$ and for $\lambda = 3$ we get $y = x$, so the eigenvector associated is of the form $(1, 1)$.

Note that the eigenvectors are orthogonal. We can verify this by calculating the inner product between them and seeing that it is equal to zero. Now it is convenient to have a unit vector (a vector of length 1) in the same direction of the eigenvectors. This is done by dividing the vectors by their respective norm. Therefore the orthonormal vectors are $(\frac{-1}{\sqrt{2}}, \frac{1}{\sqrt{2}})$ and $(\frac{1}{\sqrt{2}}, \frac{1}{\sqrt{2}})$, and the matrix of eigenvalues looks like this:

$$Q = \begin{bmatrix} \frac{1}{\sqrt{2}} & \frac{-1}{\sqrt{2}} \\ \frac{1}{\sqrt{2}} & \frac{1}{\sqrt{2}} \end{bmatrix}$$

Now this is the part where the spectral decomposition comes into play: The decomposed generating matrix A can be expressed as a diagonal matrix D:

$$QAQ^{-1} = D \tag{7.5}$$

And we know that if Q is orthonormal matrix, then its inverse Q^{-1} is its transposed matrix Q^T. Then in our specific example we have:

$$\begin{bmatrix} \frac{1}{\sqrt{2}} & \frac{-1}{\sqrt{2}} \\ \frac{1}{\sqrt{2}} & \frac{1}{\sqrt{2}} \end{bmatrix} \begin{bmatrix} 5 & -2 \\ -2 & 5 \end{bmatrix} \begin{bmatrix} \frac{1}{\sqrt{2}} & \frac{1}{\sqrt{2}} \\ \frac{-1}{\sqrt{2}} & \frac{1}{\sqrt{2}} \end{bmatrix}$$

$$= \begin{bmatrix} 7 & 0 \\ 0 & 3 \end{bmatrix}$$

Note that the diagonal matrix has the eigenvalues as the diagonal. So if we substitute this new matrix in the ellipse equation we would eliminate the xy term because of the zeroes outside the diagonal:

$$[x, y] \begin{bmatrix} 7 & 0 \\ 0 & 3 \end{bmatrix} \begin{bmatrix} x \\ y \end{bmatrix} = [7x \quad 3y] \begin{bmatrix} x \\ y \end{bmatrix}$$

giving us the final result of:

$$7x^2 + 3y^2 = 1 \qquad (7.7)$$

And this is an ellipse centered at the origin with standard equation:

$$\frac{x^2}{(\frac{\sqrt{7}}{7})^2} + \frac{y^2}{(\frac{\sqrt{3}}{3})^2} = 1 \qquad (7.8)$$

Now we can apply our knowledge of basic analytic geometry and get all the elements of the ellipse.

The applications of eigenvectors and eigenvalues spread in many areas, from quantum mechanics to Fourier transform, to analysis of Banach spaces and so on.

But I would like to emphasize that there are many applications of spectral theory to the field of decision science, starting with the classic problem of principal component analysis (PCA), which uses eigenvectors to point in the direction of the larger amount of data, which will be the same as the variance of the direction.

Another major topic in data science is the one linked to deep learning algorithms[4]: this theory has given us the chance to code in some way our geometric intuition (as Poincaré intended) into modeling, including training models and testing them.

The key concept is to create learning algorithms using geometric representations and invariants and in order to do that, we use eigenvalues, which are those values for which the following holds:

$$T\bar{x} = \lambda\bar{x} \qquad (7.9)$$

In other words, that the linear transformation applied to a vector $\bar{x}$ equals the scalar multiple of that vector. We call that vector the eigenvector.

In this section I will discuss and analyze the spectrum, which is the set of all eigenvalues of the unitary operators of the dynamical systems T^n, that are the operators induced by the automorphisms T that preserve measure.

---

[4]Deep learning algorithms are neural networks with a lot of hidden layers. In theory deep learning algorithms are those with three or more hidden layers, but I would say that this is another relative term like "big data".

7.1 Ergodic, Spectral and Functional Analyses

In order to do this, we need to establish the following definitions which are in the area of functional analysis:

Definition 81. *Adjoint Operator. We say that T^* is an adjoint operator of T if:*

$$\langle T(x), y \rangle = \langle x, T^*(y) \rangle \qquad (7.10)$$

for all $x, y \in X$ being $\langle , \rangle$ is the usual inner product in $\mathbb{R}^n$.

Definition 82. *Operator Types. We say that an operator T is:*

1. *Self-adjoint, if $T = T^*$*

2. *Normal, if $TT^* = T^*T$*

3. *Unitary, if $TT^* = T^*T = I$*

If T is self-adjoint, then $\langle x, \bar{T}(x) \rangle = \langle T(x), x \rangle = \langle x, T(x) \rangle$

Two important results of spectral theory are:

If U is unitary, and $\lambda \in \sigma(U)$ then $|\lambda| = 1$

And if U is self-adjoint and $\lambda \in \sigma(U)$, then $\lambda \in \mathbb{R}$, where $\sigma(U)$ is the set of eigenvalues of U.

For example, we will calculate the eigenvalues of the transformation T of the previous example with the torus:

$$\begin{bmatrix} 2 & 1 \\ 1 & 1 \end{bmatrix} \begin{bmatrix} x \\ y \end{bmatrix} = \begin{bmatrix} 2x & y \\ x & y \end{bmatrix}$$

One of the main objectives of this book is centered in the ergodic systems. The following theorem links spectral theory with ergodic systems that we saw in the previous sections. In particular with the rotations of the circle group.

Theorem 20. *Eigenvalue Theorem. A transformation T that preserves measure and is invertible in a finite measure space is ergodic if and only if 1 (number 1) is a simple proper value of the unitary operator U.*

Remember that an eigenvalue λ of A is called simple if its algebraic multiplicity $mA(\lambda) = 1$.

Also remember that an invertible linear transformation is a map between vector spaces and with an inverse map which is also a linear transformation.

In our case, if T is ergodic, then the absolute value of each eigenfunction of U is constant, and each eigenvalue is simple and the set of eigenvalues of U is a subgroup of the circle group.

Proof. Given that the space has finite measure, each constant function f is in L^2; given that $Uf = f$, 1 is always an eigenvalue for U. Given that the set of all constant functions in a one-dimensional space of L^2, and given that T is ergodic if and only if the only invariant functions of L^2 are constant, the first affirmation of the theorem is proven.

Remember that a function in L^2 is invariant if and only if it is the eigenfunction of U with eigenvalue 1.

Given that U is unitary, each eigenvalue of U has absolute value 1. It follows that if f is an eigenfunction with eigenvalue c, i.e. $fT(x)) = cf(x)$ almost everywhere, then $|f|$ is invariant: the ergodicity of T implies then that $|f|$ is constant.

If f and g are eigenfunctions with eigenvalue c, then f/g is an invariant function, thus g is a constant multiple of f.

Note that given that $|g|$ is a constant different than zero, f/g makes sense and is valid. This proves the simplicity of each eigenvalue.

Finally if b and c are eigenvalues of U with corresponding eigenfunctions f and g, then f/g is an eigenfunction of U with eigenvalue b/c; this proves that the eigenvalues of U form a group.

■

Ergodic theory can be studied on three different levels that can be described adequately with the words 1) algebraically, 2) geometrically and 3) analytical

The geometric level refers to the transformations in a measure space. These transformations are linked to topological spaces too.

The analytical is the study of the linear operators induced by the transformations in several L^2 spaces.

Finally the algebraic, studies the groups of automorphisms of certain Boolean algebras.

Many difficulties encountered in measure theory lie in the measure zero sets. The algebraic treatment of this difficulty is tackled considering subsets modulo measure-zero. This creates an equivalence relation between them. This is exactly the application in higher mathematics of the abstraction made by Martin Weis on the creation of equivalence classes to balance the accordion theory.

Suppose that X is a measure space with a normalized measure μ and let B be the set of equivalence classes of measurable sets, where measurable sets E and F are equivalent if and only if its symmetric difference $E \triangle F$ is zero. In general, Boolean operations work on the partitions corresponding to the equivalence relation.

The set B is a Boolean algebra under the natural Boolean operations. Indeed, (E_1, E_2) and (F_1, F_2) are pairs of equivalent sets, then $E_1 \cup F_1$ is equivalent to $E2 \cup F_2$; from this it follows that the union of two equivalence classes can be uniquely defined by selecting representatives of each class and creating the equivalence class of its union.

The same happens in the case of intersections and complements. And given that the measure is denumerably additive, it is true that unions and intersections are denumerable too.

The element zero of the Boolean algebra $\mathcal{B}$ is the class of all sets of measure zero.

Given that $\mu(E \triangle F) = 0$ then $\mu(E) = \mu(F)$, the function f can be considered defined in B. The only measure zero element in B is the zero element, in the same way the only element with measure one is the unitary element.

A structure (B, μ) i.e. a Boolean sigma-algebra with positive normalized measure is called a measure algebra.

7.2   Algebra and Geometry

The concept of measure algebra is the algebraic substitute to the geometric concept of measure space.

A transformation T that preserves measure on X induces in a natural way a mapping from B onto itself. The image of an equivalence class under the mapping is defined by selecting a representative E and creating an equivalence class $T^{-1}(E)$; the character of T that preserves measure implies that the class of the image is uniquely determined by the process and the measure of the class is the same as the original measure.

The mapping that preserves measure from B onto itself is denoted by T^{-1}. Such mapping preserves all the Boolean operations and it is an isomorphism of B onto itself.

As a note aside: A necessary and sufficient condition for B to be an automorphism of B onto itself is that the transformation T should be invertible.

A relevant question in this book is whether two transformations S and T that preserve measure are essentially the same. There are three possible answers: If S and T are taken as transformations in a measure space X, then the most adequate answer is that there is an invertible transformation $Q \in X$ that preserves measure such that $S = Q^{-1}TQ$; in this case we say that S and T are geometrically similar.

Another way to express this is: Let $T : X \to X$ and $S : Y \to Y$ two arbitrary dynamical systems, we say that T is similar to S if there is a bijection $Q : X \cong Y$ such that the following diagram commutes:

$$
\begin{array}{ccc}
X & \xrightarrow{\ T\ } & X \\
\downarrow{\scriptstyle Q} & & \downarrow{\scriptstyle Q} \\
Y & \xrightarrow{\ S\ } & Y
\end{array}
$$

Similarity in linear algebra i.e. A is similar to D if and only if there is an α with $D = \alpha A \alpha^{-1}$ is also the dynamic similarity. Intuitively, if S and T are similar then they represent the same dynamic.

If S and T are automorphisms in a measure algebra B, then the most adequate answer is that there is an automorphism Q in the algebra such that $S = Q^{-1}TQ$; in this case we say that S and T are algebraically conjugate.

Another way to put this concept is shown in this commutative diagram:

$$
\begin{array}{ccc}
Z_X & \xrightarrow{\;T\;} & Z_X \\
\downarrow{\scriptstyle Q} & & \downarrow{\scriptstyle Q} \\
Z_Y & \xrightarrow{\;S\;} & Z_Y
\end{array}
$$

where Z_X and Z_Y are measure algebras.

Finally, if S and T are unitary operators in a Hilbert space H, then the most adequate answer is that there is a unitary operator $Q \in H$ such that $S = Q^{-1}TQ$; in this case we say that S and T are spectrally equivalent.

Once again, is $T : X \to X$ and $S : Y \to Y$ and if $Q : L^2(X) \cong L^2(Y)$ then if the following diagram commutes, we say that S and T are spectrally equivalent.

$$
\begin{array}{ccc}
L^2(X) & \xrightarrow{\;T\;} & L^2(X) \\
\downarrow{\scriptstyle Q} & & \downarrow{\scriptstyle Q} \\
L^2(Y) & \xrightarrow{\;S\;} & L^2(Y)
\end{array}
$$

Definition 83. *Discrete Spectrum. We say that a transformation T has a discrete spectrum or pure punctual spectrum if there is a basis $\{f_1\}$ in L^2 i.e. an orthonormal complete set, where each element is an eigenvector of the induced unitary operator U.*

Theorem   21.   *Discrete   Spectrum   Theorem. Two ergodic transformations with discrete spectrum are conjugate if and only if their induced unitary operators are equivalent.*

Proof. It would be enough to prove that equivalence implies conjugation, so let the two given transformations be S and T, and let the induced operators U and V. And let C be the set of all the eigenvalues of U; given that U and V are equivalent, then C is also the set of eigenvalues of V.

Now, for each $c \in C$ there is a corresponding eigenvector f_c belonging to U. The eigenvalue theorem (mentioned before in this section) implies that $\mid f_c \mid$ is a constant; and without loss of generality we can assume that $\mid f_c \mid = 1$.

The eigenvalue theorem also implies (given that S is ergodic) that f_c is now uniquely determined within a constant factor of absolute value equal to 1.

The fact that U is a discrete spectrum implies that the family $\{f_c\}$ is a basis for L^2.

∎

Theorem 22. *An ergodic transformation that preserves measure with discrete spectrum is conjugate to a rotation in a compact Abelian group*

Proof. Let C be the spectrum of the given transformation and let X be the group of characters of C. If $z(c) = 0 \forall c \in C$, then z is and element of X.

The rotation $T \in X$ defined by $T(z) = zz$ is a transformation that preserves measure with discrete spectrum; moreover, its spectrum is exactly C. The spectrum's discretion follows the properties of the characters of X.

They form an orthonormal complete set on the space L^2 in X; and if f_0 is one of them, then $f_0(zz)0 = f_0(z)f_0(z)$, hence f_0 is an eigenfunction with eigenvalue $(f_0(z)$.
/newpage This argument also shows that the spectrum of T is the set of all $f_0(z)$, each one with multiplicity equal to the number of characters f belonging to X for each $f(z) = f_0(z)$.

If for each $c \in C$ we make the function $f \in X$ defined by $f_c(x) = x(c)$ correspond, then the correspondence is an isomorphism of C in the group of characters of X.

Given that $f_z(z) = c$ for each c, it follows that the spectrum of T is C. Given that the same relation shows that each element of C has multiplicity 1 in the spectrum of T, the rotation T is ergodic.

∎

As final notes of this chapter, a) a technical one: any subgroup of the circle is the spectrum of an ergodic transformation that preserves measure with discrete spectrum.

and b) an historical one: firstly, the etymology of the word *spectrum* comes from the Latin *specio* which means to look at or to view, hence the words spectacles or spectacular as something to look with or look at, in other words an image. This is more the meaning in mathematics. As stated before at the beginning of the chapter, Hilbert publishes a series of articles, where he tried to solve an integral equation containing a parameterλ, for which he uses finite quadratic forms. This shows the use eigenvalues and how eigenfunctions are orthogonal. He then links it to the principal axes as shown earlier in the chapter in the example of the ellipse widely studied in analytic geometry. The interesting further application is the characterization of eigenvalues following Poincaré's ideas. This is so huge that Hilbert leaves aside the integral equation problem.

In the 1920s John Von Neumann defines axiomatically a spectral theory for unbounded self-adjoint operators in a Hilbert space, concept I used in the book. In quantum mechanics, Bohr's proposal to characterize the hydrogen atom by associating a particle to a wave is another use of eigenvalues and the spectrum of Schrödinger's equations can be reduced to finding the energy spectrum of the Hydrogen atom to an eigenvalue problem as stated by Poincaré.

As you can see the applications of spectral theory are vast and keep growing with data science, hence the need to really understand them in order to grown and develop knowledge and apply these concepts appropriately in the field of decision science.

Chapter 8

Final Remarks

Chaos: it has no plural.

_____

The Death of Artemio Cruz
Carlos Fuentes

In this chapter I will only outline some of the main results exposed in this book. I will also summarize important concepts that are crucial in the applied and modeling fields.

Following the accordion theory as discussed with David Semach, the chapter is divided in three sections: a) the mathematical, where I will summarize all the theoretical concepts, b) the decision science one where I will mention the application of the theory and c) the philosophical one where I will outline the consequences of the theories proposed. In this way and following Martin Weis's advise, I will keep the correct balance between the parts.

8.1   Mathematics

It is important to highlight that the structure of the book was not random and never intended to explain topics separately as this style can be generally found in specialized literature. On the contrary, the book had a very well planned structure to expose these topics and concepts sin a connected way so readers and specifically decision scientists could access them with depth and contextual understanding.

So this book proposed a route from the simplest topics on homeomorphic dynamics up to spectral theory. To do this, it went through concepts from algebra, topology, measure theory, dynamical systems, differential geometry, algebraic topology, linear algebra, and functional analysis.

Throughout the book a number of topics on the mentioned areas were cross-referenced, thus giving a strong and solid foundation to algorithm-modeling and statistical analysis of even simple structures like the circle S^1.

In this way i.e. having foundational knowledge on many areas of mathematics, data and decision scientists can elevate the complexity on their own and over impose dynamical systems in it with all its possible iterations in the same way I did in this book.

One important remark is that the mathematical foundations exposed in this book are the science behind the technology (remember the discussion on science versus technology versus technique in chapter 1) and the reader must understand that what many practitioners do is merely "technique". Science is the abstraction and technique is the crafting and the "dealing with reality" in a practical way. The bridge between the two is what we call technology. When some practitioners do when they copy-paste code from libraries does not even count as technique, as it cannot be extrapolated to technology.

The increase in complexity and scale-up processes can be based on the duality of concepts e.g. the circle is not only a set of points on the plane but also a dynamic object created by iterations not axioms! i.e. an Abelian group and its rotations as explained in chapter 3.

Rotations on the circle were analyzed identifying the ergodic ones. This was done through tools like characters (which are continuous homomorphisms on the circle). This becomes clear when we highlight that given the fact that $f \in G$ (G is a group), then f is an eigenvector of the unitary operator induced by T.

Showing rotations on circles sheds light into more complex problems. Practical examples of this are analyzed in [Plata, 2020] and how they modeled people moving in a supermarket.

Characters have special importance due to the fact that they form a complete orthonormal set in L^2 and any invariant function $f \in L^2$ can be expanded in terms of its characters.

This is a clear example of how to decompose problems in real life. disassembling parts of it is not just pulling them apart as we encounter them, but paying close attention to their relational components and interactions. It is similar to the work an expert watchmaker does when taking each component of the internal mechanism of a watch to then assemble it again perfectly. The problem or model is the watch, the characters are the tools to do it and the mathematician or decision scientist is the watchmaker. Analyzing characters is a great problem solving technique and approach to technical design thinking.

Given that for unitary operators, eigenvectors with distinct eigenvalues are orthogonal, then each invariant function $f \in L^2$ (i.e. each eigenvector with eigenvalue equal to 1) is a linear combination of characters with eigenvalue 1.

Moreover, if $f \in G$ and $f(cx) = f(x)$ almost everywhere, then by continuity $F(cx) = f(x)$ everywhere and therefore $f(c^n x) = df(x)$ everywhere. This result follows when we make $x = 1$.

This is a great concept to apply in so many cases, for example when we are looking for anomalies, like fraud detection, or when analyzing autocorrelated time series like some of the stock market historical time series.

The most important results are drawn from the ergodic and spectral theorems, for example if X is a topological space and T is ergodic, then for almost all $x \in X$ the orbit of x is dense everywhere.

The rotation $T(z) = az \in S^1$ is ergodic if and only if a is not a root of the unity.

All these results are linked by the equivalence theorem which states that T is ergodic if and only if T is measurable and $(f \circ T)(x) = f(x)$, then we can say that A is ergodic if and only if the trivial character $\gamma = 1$ is the only $\gamma \in G$ that satisfies that $\gamma \circ A^n = \gamma$ for some $n > 0$.

In addition, if $A : Tor^n \to Tor^n$ is an continuous surjective endomorphism then A is ergodic if and only if the matrix $[A]$ does not have roots of unity as eigenvalues.

This concepts can be applied to any optimization problem where multiple variables are involved over time.

All of the above is reflected in the theorems of i) discrete spectrum, ii) representation, and iii) eigenvalue of spectral theory and that is why concepts of similarity conjugation and equivalence had to be introduced.

When it is said that two transformations S and T that preserve measure are essentially the same, the answer can be given in three directions: if S and T are transformations in a measurable space X, or if they are in a measurable algebra or if they are unitary operators.

This is the foundation of transferable knowledge. To apply the *Accordion Theory* discussed with David Semach is not trivial and should not be trivialized or simplified, there are several conditions (some necessary and some sufficient) to apply it. In this case, to keep balance (as suggested by Martin Weis) it is necessary to identify equivalence classes and invariants which are essential to transfer theory into practical applications.

Identifying when an object in the abstract world is the same is a great shortcut to craft solutions in the concrete world. For example see [Plata, 2006] for an example of equivalence between statistical concepts and geometry.

In summary, the mathematical concepts outlined are:

1. T is ergodic if and only if 1 is a simple eigenvalue of the induced unitary operator

2. Two transformations with discrete spectrum are conjugate if and only if its induced unitary operators are equivalent

3. An ergodic transformation that preserves measure with discrete spectrum is conjugated to a rotation in a compact Abelian group, which is the case of the circle S^1

4. We can conclude that any subgroup of the circle is the spectrum of a transformation T that preserves measure with discrete spectrum

Finally a major contribution of this book is the geometric interpretation of algebraic concepts like iterations on the circle, rotation number and ergodic transformations. This is a crucial step into problem modeling and solutions.

8.2   Decision Science

Given that data science is the field of study that overlaps mathematics and computer science to extract information and knowledge from data, and that decision science is the overlap between data science and analytics, it is fundamental to understand the three areas. What we understand by *data science* was certainly studied far before advanced computational methods were developed from the second decade of the

21st century. Clear examples come from texts like *the theory of games and economic behavior* by John Von Neumann in the 1940s and moreover algorithms like linear regression and least squares method have been in the corpus of classic mathematical theories since 1795 with the work by Gauß and Legendre on fitting astronomical data.

Higher mathematics concepts like operations in linear algebra, and eigenvector, eigenvalues, liftings, topological spaces and dynamical systems were explained in the context of applied problems. In this case investigating the asymptotic behavior of orbits of dynamical systems which are foundational in mathematical modeling.

See [Plata, 2020] for examples of mathematical concepts into data science where game theory and reinforcement learning techniques are applied to crisis management.

The *Plata-Boere third iteration recurrent theory (THIRTY)* can also be applied to data science in the field of ethical AI. The implications of a model or algorithm cannot be seen in the immediacy of the implementation, but only after iterating it several times. I discussed and analyzed practical cases about this specific application with Maxi Zattera and Daniel Scain.

The first iteration can be deceiving, bringing benefits in the short term, the second iteration can show some unforeseen consequences and the third iteration can clearly show the effects of algorithms in the long term.

Dynamical systems with ergodic transformations always have infinite orbits. The example of the circle shows that irrational rotations are ergodic.

So the link between ergodicity, irrational rotations and asymptotic behavior in statistical mechanics are closely linked.

For a data scientist who is trying to understand dynamics of processes can always refer to how orbits behave and how to analyze them[1]. Analyzing how events change in time and model them in a computational package requires some understanding of the nature of dynamical systems as shown in chapters 2 and 4.

One of the most important points of the book is to explain how homeomorphisms act on dynamical systems and how that is linked directly to the essence of neural networks. This was shown in chapters 1, 6 and 7.

I showed in the proof of the ergodic theorem a useful a application of telescopic series as a problem solving technique, which if correctly manipulated, is of the utmost use in data science and problem modeling.

Finally, links to algebraic topology show different ways to model problems as there can be homomorphisms and isomorphisms by which the problem can be simplified. It is common in data science to look for the leaner solution through these methods.

---

[1]Orbits are the first generation of information out of data as we distinguish the terms data (which is the plural of *datum*), information and knowledge.

In addition, the reader could realize that the future of data science relies on understanding these concepts as they link many theories and connect many dots injecting creativity to problem solving and other applications of data science and modeling.

Connecting the dots is also done by iterations in different spaces. Debolina Guha Majumdar suggested applied cases related to model applications, data strategy programs and responsible AI altogether. This is precisely the most powerful of all connections: the cross categorical and cross disciplinary ones.

This book will serve as a guide to see problems from the abstract realm of pure mathematics, but also will show the practical applications of the abstractions.

8.3   Philosophy

The main thesis of the book is the application of two theories proposed by the author, Alec Boere and close collaborators: David Semach, Martin Weis and Maxi Zattera to construct mathematical foundations for data science and hence find practical applications outside the pure mathematical field.

The first one, the Plata-Boere third iteration recurrent theory (THIRTY), states that we need at least three iterations or repetitions of the same function to see change. As explained throughout the book, a condition for the theory to be applicable on either practical or theoretical cases is the property of *recurrence*. Another condition is that the function must be the same.

When one applies the first step of a process or executes an action then one can only see their short term effects; when one iterates or repeats a process or action (which we call function) we start seeing the evolution of these short term effects.

These repetitions must impose a rhythm, that is why we call them iterations. Iterations in the philosophical sense are repetitions with cadence or rhythm, and they are indeed the drivers for change.

But it is not until we iterate for at least three times that we start seeing some stable patterns.

Examples in mathematics are exposed and explained in this book. Instances of this outside mathematics are numerous. In other spheres, examples are multiple: in marketing or publicity, the periodic repetition of a campaign or a commercial is key to achieve results. Lubna Ahbedin suggested many examples outside the immediate business sphere, for instance the scheduled repetition of the same song creates familiarity and acceptance and any concatenated process is another example.

Understanding politics by iterations is masterly exemplified by Kofi Awuma, who also provided me with some examples in our after-hours discussions.

In art, pointillism is clearly iterative and any spiral-wise phenomenon is iterative. When we say "it spirals out of control" we are talking about the repetition of actions in the same sense, direction and context until the system diverges and gets uncontrollable. But it needs to happen at least three times to see the "out-of-control" in the spiral.

A rather poetic application of the theory is within the concept of "return", which implies renewal and rebirth or (how Gilles Deleuze puts it) "the power of beginning and beginning again", see [Deleuze, 1994]. But the interesting application to business suggested by Alec Boere is on marketing when companies re-brand themselves or revisit their strategy every year; in this case, they are applying an iterative process, but for the process to start seeing change it must iterate at least three times.

Reinvention, the "power of failing forward" and innovation itself are clear applications of this theory. And moreover changes and innovation consequences can be fully explained through this theory (that is why it makes it a cognitive theory).

Highlighted by Markovic Ioannou, in technology, proofs of concepts which are generally the first iterations are followed by pilots which are the second iterations but in themselves they are not enough to provoke stable change; and what we have observed is that we need a third one to see meaningful change as the theory indicates.

Another example is when taking the case of any repetitive AI algorithm or technological policy which means well at the beginning but has opposite consequences after three or four iterations e.g. privacy scandals, or vital information loss due to weak policies and in general ethical AI problems.

Explanation of phenomena, is the reason why our theory is elevated to an epistemological level, which is closer to a *philosophy*. Understanding that directed iterations allow reinvention is a corollary of this theory.

Explaining the world through iterative power can be found even in the personal and biological levels. It is important to note that explaining general situations and effects in day-to-day life, adds an existential dimension to the theory.

As exposed throughout the book, iterations depend on functions, but understanding these functions in isolation is not enough to understand the system of iterations (or dynamical system) and moreover predict the asymptotic behavior or long term effect of rhythmic repetitions.

But how do we monitor these effects of repetitions? well, to assess the long term effects of recurrent iterations we need the *accordion theory*; and this completes the cycle.

Iterations of abstract functions should be evaluated on the practical level and as Martin Weis puts it, "balance is key to understand abstractions and to land change".

This balance is the artful application of "abstract and concrete" and "general and particular". Like an accordion one needs to stretch to the theoretical and squeeze into the practical within the same functional applications.

The theory also states that if the movement between approaches introduces a different function, then it would not be possible to neither land change, nor explain it and let alone improve. As put by Kofi Awuma "consistency is essential".

The theory also requires the identification of equivalent classes to iterate between them and not between the elements of the functions. Practical and abstract are not *parts* but a *whole*.

In other words, the accordion theory acts on classes not elements and moreover the balance is achieved by the same function and its inverse[2].

Examples in change management are vast, for example the formation of virtuous circles when we monitor the number of repeated behaviors and establish the cadence and pattern of a new process only to go back to the drawing board (the theory) and tweak the process to manage change efficiently.

As Martin Weis mentioned in one of our discussions in the context of strategy "it is not only the number of times you repeat, but how you jump in and out between repetitions".

That is why in the book I construct theories by going in and out of theoretical and practical applications and and zoom in and out from the general to a particular with examples. And as Maxi Zattera mentioned to me in our discussions, "tactical solutions are also crucial".

In both science and technology, models and solution for business and moreover, evaluation of current and future processes are mainly understood through the *accordion theory*. If we go on with theoretical models or pure data science models, the value will be lost. It is in

---

[2]The theory also requires the existence of an inverse or conjugate to be able to move back and forth between them like an accordion.

the actual recommendation, the actions behind the analysis and the actual decisions made based on data-led models that change is produced. Iterations between the theory and practice must be artfully combined to achieve transformations.

Finally the combination of the two theories served as a basis for:

a) construct the concept of probability outside the classic axiomatic approach. This was rather done by iterations on basic functions i.e. dynamical systems a circle is not the set of points that are equidistant to another one called center, but a circle is a dense orbit of an irrational rotation iterating infinitely. In this same way, probability is not the division of two numbers, the positive cases divided by the possible cases, but the number of times a dynamical system recurs on a set. The probability is precisely the size of the set.

b) a pedagogical proposition to understand mathematical foundations of data science models. The link between many fields of mathematics: abstract algebra, group theory, algebraic topology, differential topology, spectral theory, linear algebra, complex variable, measure theory, dynamical systems and ergodic theory rather than a specialized one proposes a wider and better understanding of the practical applications and moreover, the foundations of data science methods

and therefore c) shed light towards an approach to R&D, increase knowledge in the field of decision science and practically land change more effectively in data science, AI and technology.

Chapter 9

Recommended and Cited Bibliography

1. Arfken G., *Mathematical Methods for Physicists: A Comprehensive Guide*, Academic Press; 7th edition, Amsterdam, 2012

2. Arnold V.I., *Ergodic Problems of Classical Mechanics*, W.A. Benjamin Inc., New York, 1968

3. Bateson G., *Steps to an Ecology of the Mind*, Chandler Publishing Company, Toronto, 1972

4. Bergson H., *Creative Evolution*, Translated by Arthur Miychell, Random House, New York, 1944

5. Berlanga R., *Simetrías y Coloraciones*, Misc Mat., 22 3-21, 1995

6. Berlanga R., *A Mapping Theorem for Topological Sigma-Compact Manifolds*, Compositio Mathematica 63 209-216, 1987

7. Berlanga R. *Measures on Sigma-Compact Manifolds and Their Equivalence Under Homeomorphisms*, J London Math Soc 27 63-74, 1983

8. Bisewski K., Ivanovs, J., *Zooming-in on a Lévy process: failure to observe threshold exceedance over a dense grid*, Electron. J. Probab. 25 (2020), article no. 113, 1–33. ISSN: 1083-6489 https://doi.org/10.1214/20-EJP513

9. Bisewski K., Debicki K., Rolski T., *Derivatives of sup-functionals of fractional Brownian motion evaluated at $H = \frac{1}{2}$*, Electron. J. Probab. 27 (2022), article no. 129, 1–35. ISSN: 1083-6489 available at https://doi.org/10.1214/22-EJP848

10. Bisewski K., Asnovidov, G., *On the Speed of Convergence of Discrete Pickands Constants to Continuous Ones*, arXiv preprint arXiv:2108.00756 (2022)

11. Blanchard P., *Differential Equations*, Brooks/Cole Publishing Co., Boston, 1998

12. Boere A., Plata S., *Can Tree-Search Algorithms be a Game Changer in Banking?*, Infosys Consulting, Banking, https://www.infosysconsultinginsights.com, January 18 2021

13. Bohr N., *The Philosophical Writings*, Ox Bow Pr, Oxford, 1987

14. Branner B., *Real and Complex Dynamical Systems*, Kuwer Academic Publishers, Amsterdam, 1993

15. Calvino I., *Invisible Cities*, Vintage Classics, 1st edition, New York, 1997

16. Canaparo C., *Geo-Epistemology*, Peter Lang, Oxford, 2010

17. Canaparo C., *La Cuestión Periférica Heidegger, Derrida, Europa*, Peter Lang, 2021

18. Cornfield I.P., *Ergodic Theory*, Moscow, Springer Verlag, 1982

19. Crichton M., *The Lost World*, Knopf Publishers, New York, 1995

20. Christensen C., *Course Research: Using the Case Method to Build and Teach Management Theory*, Academy of Management Learning Education, Vol. 8, No. 2, 240 – 251, 2009

21. Deleuze, G., *Difference and Repetition*, Trans. by Paul Patton, Athlone Press, London, 1994

22. Descartes R., *Discourse on Method*, Hackett Publishing Company 3rd edition, New York, 1998

23. Devaney R., emphAn Introduction to Chaotic Dynamical Systems, CRC Press 2nd edition, 2003

24. Falorsi L., et al. *Explorations in Homeomorphic Variational Auto-Encoding.* ArXiv Preprint ArXiv:1807.04689, 2018

25. Friedrichs K.O., *Spectral Theory of Operators in Hilbert Spaces*, Springer Verlag, New York, 1980

26. Fuentes C., *La Muerte de Artemio Cruz*, Fondo de Cultura Economica, Mexico City, 1962

27. Guillemin V., Pallack A., *Differential Topology*, American Mathematical Society Reprint edition, Providence Ri, 2010

28. Gurney K., *An Introduction to Neural Networks*, UCL Press Limited, London, 1997

29. Greenberg M.J., *Algebraic Topology: A First Course*, Benjamin/Cummings Publishing Co. Madison, 1981

30. Halmos P., *Lectures in Ergodic Theory*, Chelsea Publishing Company, Boston, 1956

31. Hamdy T., *Operations Research: An Introduction*, Pearson; 10th edition, London, 2017

32. Helmberg G., *Introduction to Spectral Theory in Hilbert Spaces*, North-Holland Publishing Co. London, 1969

33. Hu S-T., *Elements of General Topology*, Holden-Day, Inc, Toronto, 1965

34. Kundera M., *Slowness*, Faber & Faber, London, 1997

35. Loomis L., *Advanced Calculus*, Jons and Bartlett Publishers Inc, Boston Ma. 1990

36. Nachbin L., *The Haar Integral*, D. Van Nostrand Co., Boston, 1965

37. Nadkarni M.G., *Basic Ergodic Theory*, Birkhauser Verlag, New York, 1995

38. McCulloch W., Pitts W., *A Logical Calculus of the Ideas Immanent in Nervous Activity*, Bulletin of Mathematical Biophysics, vol. 5, pp. 115–133, 1943

39. Percival I., *Introduction to Dynamics*, Cambridge University Press, Cambridge, 1985

40. Petersen K., *Ergodic Theory*, Cambridge University Press, Cambridge, 1983

41. Plata S., *A Note of Fishers Correlation Coefficient*, Applied Mathematics Letters, Volume 19, Issue 6, Pages 499-502, June 2006

42. Plata S., *Visions of Applied Mathematics*, Peter Lang, Oxford, 2008

43. Plata S., *Editorial*, Mathematics Today, Feb 1, 2020

44. Plata S., Sarma S., Lancelot M., Bagrova K., Romano-Critchley D., *Simulating Human Interactions in Supermarkets to Measure the Risk of COVID-19 Contagion at Scale* , arxiv.org/abs/2006.15213, 2020

45. Plata S., Sarma S., *Data-Centric Crisis Management*, Mathematics Today, Pages 144-146, August 2020

46. Plata S., *Leading Data Science Teams Through Gamification: 4 Principles that Organizations Can Borrow from Games*, Infosys Consulting, https://www.infosysconsultinginsights.com, 2021

47. Plata S., Raczkiewicz M., Sun H. *A new Approach to Forecast Accuracy*, AI & Automation Business Journal, November Issue (2022), pp. 52-65 available at https://www.infosysconsultinginsights.com/insights/ai-automation-business-journal

48. Plata S., Hays C., *Machine Learning to Better Identify Product & Market Synergies*, AI & Automation Business Journal, November Issue (2022), pp. 37-51 available at https://www.infosysconsultinginsights.com/insights/ai-automation-business-journal

49. Poincaré, H., *Science and Method*, T. Nelson Publisher, London, 1914

50. Radjavi H., *Spectral Theory*, Springer Verlag, New York, 1970

51. Raissi,M., Perdikaris P., Karniadakis G.E., *Physics-Informed Neural Networks: A Deep Learning Framework for Solving Forward and Inverse Problems Involving Nonlinear Partial Differential Equations*, Journal of Computational Physics, S0021-9991(18)30712-5, November 2018

52. Rosenblatt F., *The Perceptron: A Probabilistic Model for Information Storage and Organization in the Brain*, Psychological Review, vol. 65, no. 6, p. 386–408, 1958

53. Royden H.L. *Real Analysis*, The Macmillan Company, Boston, 1968

54. Rudin W., *Principles of Mathematical Analysis*, McGraw Hill, New York, 1976

55. Rudin W., *Functional Analysis*, McGraw Hill, New York, 1991

56. Rudin W., *Real and Complex Analysis*, McGraw Hill, New York, 1974

57. Simondon G., *On the Mode of Existence of Technical Objects*, translated by Cecile Malaspina and John Rogove, Univocal Publishing, Minneapolis, 2017

58. Sinai Y. G., *Topics in Ergodic Theory*, Princeton University Press, Princeton N.J., 1994

59. Spivak M., *Calculus*, Cambridge University Press, Cambridge, 2006

60. Spivak M., *Calculus on Manifolds*, Addison-Wesley Publishing Company, New York, 1965

61. Spivak M., *A Comprehensive Introduction to Differential Geometry Vol. 4*, Publish or Perish; 3rd edition, 1999

62. Thiede E.H., Zhou W., Kondor R., *Autobahn: Automorphism-based Graph Neural Nets*, 35th Conference on Neural Information Processing Systems (NeurIPS 2021)

63. Venkov A.B., *Spectral Theory of Automorphic Functions and Its Applications*, Kluwer Academis Publishers, Amsterdam, 1990

64. Von Neumann J., Morgenstern O., *Theory of Games and Economic Behavior*, Princeton: Princeton University Press, 1947

65. Walters P., *An Introduction to Ergodic Theory*, Springer Verlag, Berlin, 1982

66. Waltershausen Von S., *Gauß zum Gedächtniss: Biographie Carl Friedrich Gauß*, Gutenbergplatz Leipzig, 2012

67. Wiener N., *Cybernetics or Control and Communication in the Animal and the Machine*, The MIT Press, Cambridge Massachusetts, 1965

Index

www.ingramcontent.com/pod-product-compliance
Lightning Source LLC
Chambersburg PA
CBHW051753200326

41597CB00025B/4538